T0275728

Experimental Design in Petroleum Reservoir Studies

Experimental Design in Petroleum Reservoir Studies

Mohammad Jamshidnezhad
Senior Reservoir Engineer, NISOC

AMSTERDAM • BOSTON • HEIDELBERG • LONDON
NEW YORK • OXFORD • PARIS • SAN DIEGO
SAN FRANCISCO • SINGAPORE • SYDNEY • TOKYO

Gulf Professional Publishing is an imprint of Elsevier

Gulf Professional Publishing is an imprint of Elsevier
225 Wyman Street, Waltham, MA 02451, USA
The Boulevard, Langford Lane, Kidlington, Oxford, OX5 1GB, UK

Notices
Knowledge and best practice in this field are constantly changing. As new research
and experience broaden our understanding, changes in research methods, professional
practices, or medical treatment may become necessary.

Practitioners and researchers must always rely on their own experience and knowledge
in evaluating and using any information, methods, compounds, or experiments described
herein. In using such information or methods they should be mindful of their own safety
and the safety of others, including parties for whom they have a professional responsibility.

To the fullest extent of the law, neither the Publisher nor the authors, contributors, or editors,
assume any liability for any injury and/or damage to persons or property as a matter of
products liability, negligence or otherwise, or from any use or operation of any methods,
products, instructions, or ideas contained in the material herein.

ISBN: 978-0-12-803070-7

Library of Congress Cataloging-in-Publication Data
A catalog record for this book is available from the Library of Congress

British Library Cataloguing-in-Publication Data
A catalogue record for this book is available from the British Library

For information on all Gulf Professional Publishing publications
visit our website at http://store.elsevier.com/

This book has been manufactured using Print On Demand technology. Each copy is produced
to order and is limited to black ink. The online version of this book will show color figures
where appropriate.

Working together
to grow libraries in
developing countries

www.elsevier.com • www.bookaid.org

Contents

Biography		**vii**
Preface		**ix**
1	**Introduction**	**1**
	1.1 Petroleum reservoirs	1
	1.2 Petroleum rock properties	1
	1.3 Volumetric calculations in a reservoir	3
	1.4 Reservoir heterogeneity	4
	1.5 Reservoir models	6
	1.6 Experimental design	8
2	**Reservoir modeling**	**9**
	2.1 Introduction	9
	2.2 Sources of data for reservoir modeling	10
	2.3 Reservoir characterization	10
	2.4 Mathematical modeling	44
	2.5 Model verification	55
3	**Experimental design in reservoir engineering**	**57**
	3.1 Introduction	57
	3.2 Errors in mathematical modeling	57
	3.3 Uncertainty in reservoir data	61
	3.4 Uncertainty analysis	63
	3.5 Experimental design	74
4	**Case studies**	**99**
	4.1 Introduction	99
	4.2 Case study 1	99
	4.3 Case study 2	105
	4.4 Case study 3	134
	4.5 Case study 4	142
	4.6 Case study 5	147
	4.7 Case study 6	156
Appendix: F distribution values		**169**
References		**171**
Index		**175**

Biography

Mohammad Jamshidnezhad is a Senior Reservoir Engineer who has worked (in Canada and Iran) for more than 14 years on carbonate and sandstone reservoirs and become specialized in reservoir engineering aspects of enhanced oil recovery (EOR), PVT, reservoir simulation and uncertainty assessments. He holds a PhD degree in chemical engineering from the University of Tehran (Iran). Mohammad was a Research Scholar in the Department of Petroleum Engineering at Curtin University of Technology in Australia in 2003 where he performed successful experiments on Enhanced Oil Recovery. He was employed as a postdoctoral researcher working on a simultaneous water and gas injection project at the Petroleum Engineering Section of the Department of Geoscience and Engineering at Delft University of Technology (The Netherlands) during 2007−2008. Mohammad has taught industry short courses in reservoir simulation, PVT, and uncertainty analysis. He is the author of several peer-reviewed journal and international conference papers.

Preface

Petroleum reservoir engineering is one of the most attractive fields at universities and colleges since most graduates find good positions in petroleum companies and work in different disciplines like estimating hydrocarbon in-place, reserve, best enhanced oil recovery methods, and reservoir management.

One of the main duties for reservoir engineers is reservoir study, which starts when a reservoir is explored and continues until reservoir abandonment. Reservoir study is a continual process, which means after a period of production the study should be updated.

A reservoir study starts with reservoir characterization: that is, gathering data (geological, geophysical, drilling and production) and building a geological model. The geological model (fine model) is upscaled and then initialized subject to initial conditions of the reservoir. The dynamic model is run by a reservoir simulator and the model results are then compared with observed (field) data. This step of reservoir study is called history matching. If the comparisons are reasonable, the model behaves similarly to the actual reservoir and it is then used for predicting the future of the reservoir. Because of reservoir complexity and limited information and data, the reservoir characterization is not conducted completely and precisely. This means that uncertainties always exist in reservoir characterization and reservoir engineers cannot define (model) a reservoir deterministically. Uncertainty in reservoir modeling causes difficulties in reasonable history matching and prediction phases of study. Quantifying and analyzing uncertainties could relieve the difficulties.

The focus of this book is on experimental design to analyze and to quantify these uncertainties. The book is divided into four chapters. In Chapter 1, an introduction to petroleum reservoirs is presented. Reservoir modeling is discussed in Chapter 2. In Chapter 3, uncertainties in reservoir modeling and experimental design methods are explained, and finally six case studies are discussed in Chapter 4. Five cases are run using black-oil reservoir simulators, and a thermal reservoir simulator is used for the sixth case. The explained case studies cover a wide range of reservoir studies: two conventional petroleum reservoirs, a fractured carbonate reservoir, steam-assisted gravity drainage (SAGD) in a heavy oil reservoir, miscible water alternate gas (WAG) into a reservoir, and a hydraulically fractured shale-gas reservoir.

Last but not least, I would like to thank Prof. Mehran Pooladi-Darvish (Fekete Associate Inc.) for initiating the concept behind this book. I would also like to thank all Senior Reservoir Engineers at the Petroleum Engineering Department of National Iranian South Oil Company (NISOC) for providing some data of the second case study. Finally, I appreciate the editorial comments provided by Dr. Alireza Jamshidnejad.

Mohammad Jamshidnezhad
Senior Reservoir Engineer
September 2014

Introduction

1.1 Petroleum reservoirs

Most petroleum geologists believe that crude oil is a result of diagenesis (process of rock compaction that leads to change in physical and chemical properties of the rock) of buried organic materials and therefore is a vital characteristic of sedimentary rocks, not of volcanic rocks. Based on the views of petroleum geologists, the following five conditions generate petroleum traps [Selley, 1998]:

1. Rich source rocks of organic materials that produce hydrocarbons;
2. Source rock heated sufficiently to liberate crude oil;
3. A reservoir to collect liberated hydrocarbons, which should be porous and permeable enough to store and transfer hydrocarbons;
4. This reservoir must have impermeable cap rock to prevent hydrocarbon escaping to the surface;
5. Source rock, reservoir and cap rock should be arranged in a way that enables the trapping of hydrocarbons.

Theoretically, any sedimentary rock could be a hydrocarbon reservoir; however, in reality only sandstone (clastic sedimentary rocks composed mainly of quartz) and carbonate (rocks composed of calcite or dolomite) rocks are the main hydrocarbon resources in the world. There are also shale (a fine-grained rock composed of clay) oil and gas fields in some regions.

In contrast to sandstone reservoirs, carbonate reservoirs are sensitive to diagenesis processes and their reservoir qualities strongly depend on certain factors carried out during the diagenesis. Diagenesis processes in carbonate rocks cause fracturing, dolomitization, dissolution and cementation. Some of these processes, such as fracturing and dolomitization, improve reservoir qualities. In general, sandstone reservoirs have higher reservoir qualities in comparison to carbonate and shale reservoirs.

1.2 Petroleum rock properties

Different properties of reservoir rocks are characterized when a petroleum reservoir is studied. These properties are: mineralogy, grain size, porosity, permeability, acoustic properties, electrical properties, radioactive properties, magnetic properties, and mechanical properties.

Mineralogy and grain size: Quartz and calcite are the most common minerals in reservoir rocks. Trace minerals are often present as individual grains or as cement. Grain size and sorting can vary considerably; however, reservoir quality tends to decrease with decreased grain size. Accordingly, very finely grained rocks (such as shale) tend to have sealing properties.

Experimental Design in Petroleum Reservoir Studies. DOI: http://dx.doi.org/10.1016/B978-0-12-803070-7.00001-6

Acoustic properties: Acoustic measurements include sonic and ultrasonic ranges. The primary and most routine use of acoustic measurement in reservoir engineering is porosity determination.

Electrical properties: Studies of electrical properties in rocks are mainly performed for determination of formation resistivity and water saturation.

Radioactive properties: Geological age of a formation and the volume of shale in the formation are estimated by measuring radioactivity in rocks. The gamma logs (a tool for measuring the natural radioactivity from potassium, thorium, and uranium isotopes in the earth) are used as a shaliness indicator in petroleum reservoir studies.

Magnetic properties: Nuclear magnetic resonance (NMR), a subcategory of electromagnetic logging, measures the induced magnetic moment of hydrogen nuclei contained within the fluid-filled pore space of porous media (reservoir rocks). NMR provides information about: the volume (porosity) and distribution (permeability) of the rock pores, rock composition, and type and quantity of fluid hydrocarbons.

Mechanical properties: Mechanical properties of rocks are important in formation evaluation, drilling, development planning and production. These properties are useful in borehole stability analysis, sand production prediction, hydraulic fracture design and optimization, compaction/subsidence studies, drill bit selection, casing point selection and casing design.

Porosity: Usually petroleum rock pores are filled with connate water and hydrocarbons. Porosity is the ratio of pore volume to bulk volume and is usually reported as percentage. Two porosity values are usually measured: total porosity and effective porosity. Total porosity is the fraction of rock bulk volume that is void, whether the individual pores are interconnected or not. Effective porosity is the ratio of connected void space to rock bulk volume. It is the effective porosity that reservoir engineers are interested in and, in almost all cases, the porosity measured in the laboratory is the effective porosity.

Uniformity of grain size, degree of cementation, amount of compaction during and after deposition, and methods of packing are the factors governing the magnitude of porosity [Tiab and Donaldson, 2004].

Porosities are measured in core laboratories, as well as by using the sonic-acoustic log, the formation density log, and the neutron porosity log [Tiab and Donaldson, 2004]: In the core laboratory bulk volume, pore volume, rock matrix volume, and irreducible water saturation are measured. By knowing these parameters, total porosity and effective porosity are calculated. Commonly, mercury injection and gas compression/expansion are used to determine total porosity and effective porosity, respectively.

In the sonic log, the time required for a sound wave to travel through one foot of formation is measured. This transiting time is then correlated to the porosity.

In the formation density log, the bulk density of the reservoir rock is measured. Using bulk density, matrix density and average density of fluids filling the formation, porosity is evaluated.

The neutron log is sensitive to the amount of hydrogen atoms in a formation. In the neutron log, a neutron source is employed to measure the ratio of the concentration of hydrogen atoms in the material, to that of pure water at 75°F. This ratio (called hydrogen index, HI) is directly related to porosity.

Permeability: The second main property of a reservoir rock, after porosity, is permeability. Porous medium is not sufficient for a reservoir rock—the pores must be connected to each other. Permeability is a measure of the rock's capability to transport fluids. Primary work on permeability was done by Darcy in 1856. Darcy's law is formulated as:

$$U = \frac{k(P_1 - P_2)}{\mu L} \tag{1.1a}$$

The unit of permeability (k) is the darcy, which is the permeability of a rock transporting a fluid of 1 cp viscosity (μ) with a velocity (U) of 1 cm per second and 1 atmosphere pressure drop ($P_1 - P_2$) along a rock of 1 cm length (L). Permeability of most hydrocarbon reservoirs is much less than one darcy, so the unit of milli-darcy (0.001 darcy, abbreviated "md") is usually applied.

For laminar gas flow through porous media, Darcy's law is shown as:

$$U = k(P_1^2 - P_2^2)/(2\mu \cdot L \cdot P_{ave}) \tag{1.1b}$$

The permeability of a hydrocarbon reservoir is measured (or estimated) using one of the methods described in the following paragraphs.

The first method is by using well testing data. In a well test, by changing the flow rate of a well, variation in well bottom-hole pressure is recorded as a function of time. The flow rate of a well is changed by increasing or decreasing the rate. The pressure change is analyzed by plotting the recorded pressure and its derivative versus time. The two most common tests are the buildup and drawdown tests. In a buildup test the well is shut in after a period of production and then its pressure is measured. In a drawdown test the pressure is measured in a well that is open after a period of well shutting in.

The second method is measuring the permeability in a core laboratory. A known gas (air or nitrogen) is injected into a core (or a plug) at controlled velocity and then the pressure drop is measured. Using Darcy's law (Eq. 1.1b), permeability is calculated and it is then extrapolated to the zero value of the reciprocal pressure ($1/p$) to estimate liquid (oil or water) permeability.

In the presence of more than one fluid, permeability is referred to as the *effective permeability*. In this case, the ratio of effective permeability of any phase to the absolute permeability of the rock is called the relative permeability (k_r) of that phase. Darcy's law for multiple phase flow through the rock is formulated as:

$$Q_p = \frac{-kk_{rp}A}{\mu_p}\left(\frac{dP}{dx}\right)_p \tag{1.2}$$

where the subscript p denotes phase.

Shape and size of grains, lamination, cementation, fracturing and solution are the factors governing the magnitude of porosity [Tiab and Donaldson, 2004].

1.3 Volumetric calculations in a reservoir

In oil reservoirs, the original oil in place (OOIP) is calculated as:

$$OOIP = V_b\varphi(1 - S_{wc}) \tag{1.3}$$

where V_b is the bulk volume of reservoir rock, φ is the average porosity of the reservoir rock and S_{wc} is the average connate water of the reservoir rock.

The bulk volume, V_b, is obtained from geological, geophysical and fluid pressure analysis. The product $V_b\varphi$ is called the pore volume (PV) and is the total volume that may be occupied by fluids. Similarly, $V_b\varphi(1 - S_{wc})$ is called the hydrocarbon pore volume (HCPV) and is the total reservoir volume that may be filled with hydrocarbons.

The oil volume calculated using Eq. 1.3 is usually expressed as stock tank oil initially in place (STOIIP) as:

$$STOIIP = N = V_b\varphi(1 - S_{wc})/B_{oi} \tag{1.4}$$

where B_{oi} is the initial oil formation factor.

Although the amount of hydrocarbons inside a reservoir is a constant quantity, the reserve (i.e., recoverable oil and gas) depends on the production technique. Theoretically, the maximum amount of oil that may be removed from a reservoir is called movable oil volume, MOV, and is calculated as:

$$MOV = V_b\varphi(1 - S_{or} - S_{wc}) \tag{1.5}$$

In this equation, S_{or} is the residual oil saturation that depends on the recovery mechanism.

Three hydrocarbon recovery mechanisms are used, respectively, in a petroleum reservoir:

- Primary recovery, or natural depletion, where the volume of hydrocarbons may be produced utilizing the natural energy of the reservoir. Natural energies come from one or more of the following: gas cap expansion (gas cap drive), aquifer expansion (water drive), pore compaction, fluid expansion, solution gas.
- Secondary recovery, where the volume of recovered hydrocarbons is incremented by injecting water or immiscible gas (hydrocarbon gas such as methane or non-hydrocarbon gas such as nitrogen, carbon dioxide) into the reservoir. Alteration injection of water and immiscible gas is also considered as a secondary recovery mechanism.
- Enhanced oil recovery (EOR), where the volume of recovered hydrocarbons is raised by injecting chemicals, hot water, steam and surfactants into the reservoir. In an EOR process, chemical and/or physical properties of reservoir oil and rock may change.

1.4 Reservoir heterogeneity

In contrast to homogeneous reservoirs, geological and petrophysical properties of heterogeneous reservoirs vary with position. More variations in reservoir properties result in more heterogeneity. Reservoir properties variation means that there is no homogeneous petroleum reservoir in the real world. In other words, all petroleum reservoirs are heterogeneous but the degree of heterogeneity is different from one reservoir to another. Usually, the degree of heterogeneity in sandstone reservoirs is less than in carbonate reservoirs. Figure 1.1 shows a typical heterogeneous reservoir, where the permeability varies from 10 md to 1000 md.

Among the petroleum reservoir properties, heterogeneity of reservoir rock properties affects more than reservoir fluids. Fanchi (2010) and Schulze-Riegert and

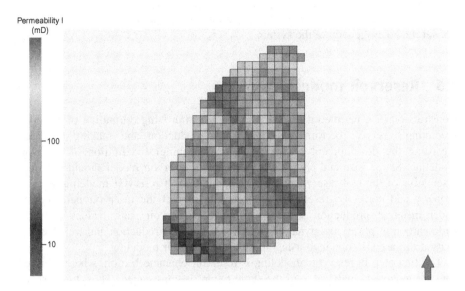

Figure 1.1 Permeability variation in a typical heterogeneous reservoir, (view from top).

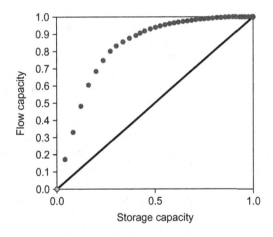

Figure 1.2 Typical Lorenz plot.

Ghedan (2007) summarized four levels of reservoir rock heterogeneities as microscopic (for the scale of $10-100\,\mu m$), macroscopic (for the scale of $1-100\,cm$), mega-scale (for the scale of $10-100\,m$) and giga-scale (for the scale of greater than $1000\,m$). It should be noted that degree of heterogeneity along the reservoir length is different from the vertical direction; some properties such as porosity and permeability have a higher degree of heterogeneity in the vertical direction.

There are a few methods to determine degree of heterogeneity. Among them, the Lorenz method, where the cumulative flow capacity, $\Sigma(kh)$, is related to the cumulative storage capacity of the reservoir, $\Sigma(\varphi h)$, is a common method. Figure 1.2 shows a typical Lorenz plot. The greater the deviations of this curve from the 45° line, the greater the heterogeneity of the system.

1.5 Reservoir models

Construction of a petroleum reservoir model for improving estimation of reserves, predicting reservoir performance, increasing production and making decisions regarding the development of the field is a meaningful definition of reservoir modeling. Shape, size and physical properties of a reservoir model should be a representative of the real reservoir that is being studied. If reservoir modeling is done properly and correctly, reservoir engineers can predict the reservoir performance under different production scenarios. Approval reservoir static values (such as hydrocarbon in-place) and dynamic conditions (such as production and well productivity drop) are also obtained using appropriate reservoir modeling.

The first step in reservoir modeling is reservoir characterization, where required data are gathered, analyzed, and the static model is constructed. Thus, to construct an appropriate reservoir model, there should be two necessities: 1) size and properties of the reservoir should be known (in reservoir modeling this phase is called reservoir characterization), and 2) equations of fluid flow through porous media should be solved accurately. In practical applications, there are some limitations to these requirements: first, no one is able to measure or examine all reservoir properties, and second, some of the rules and phenomena of fluid flow through porous media are still incomplete. Also, numerical methods to solve the equations of fluid flow have their own limitations. Among these limitations, the reservoir characterization phase has the most restrictions (or, as a more correct sentence, has the most uncertainties).

Schulze-Riegert and Ghedan (2007) mentioned three sources for uncertainties in reservoir modeling: measuring errors, mathematical errors, and incompleteness of data. In brownfields (where there are histories of pressure and fluid production), one way to overcome the errors and incompleteness is to change the reservoir properties so that field data (pressure and fluid production) and model results are matched. This method is called history matching. Once the model historically matches field data, it may behave the same as the actual reservoir under future constraints.

However, history-matching suffers from three difficulties:

- The solutions of fluid flow equations are known (we know the pressure and production data), but the input parameters (reservoir properties) are uncertain. Thus, we can say that history matching is an inverse problem and therefore could have several solutions. Figure 1.3 illustrates a history-matching case where two models with different parameters are matched with observed data.
- It is a time-consuming phase and occupies a large portion of a reservoir study time. Experience shows that normally 40% of the time for a reservoir study is spent on history-matching.
- Sometimes unrealistic reservoir properties are needed to achieve a history-matched model [Satter and Thakur, 1994]. Figure 1.4 depicts water relative permeability data needed for a case seeking a successful history-matched model.

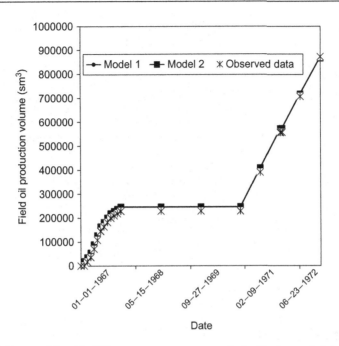

Figure 1.3 Two models with different parameters are matched with observed data.

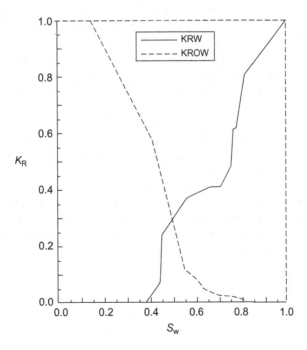

Figure 1.4 Unrealistic water relative permeability data for a successful history-matching case.

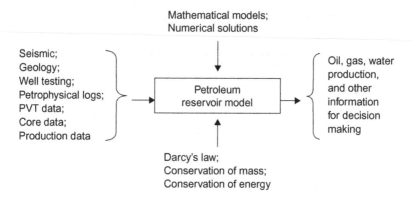

Figure 1.5 General diagram of reservoir modeling.

In the next chapters, the procedure of reservoir modeling and history-matching are thoroughly explained.

1.6   Experimental design

Experimental design is a viable tool to acquire knowledge and to optimize reservoir processes at minimum cost and time [Montgomery, 2001]. It is a statistical method to study cause-effect and phenomena-response relationships in processes and phenomena [Lazić, 2004].

As the behavior of a petroleum reservoir is controlled by the interactions of various factors, experimental design can be used to study the effects of one or more factors on reservoir performance. Figure 1.5 illustrates a general model for reservoir modeling.

By using experimental design, reservoir engineers are able to condition the reservoir and to adjust the most influential factors (parameters), so that the reservoir model is history matched, the hydrocarbon production is maximized, and the reservoir is at its optimum conditions.

This book claims that experimental design could be used to improve reserve estimation, to do history matching with fewer difficulties, to predict reservoir performance with more reliability, to increase production, and to make effective decisions regarding the development of the field. The methodology of experimental design is explained in detail in Chapter 3.

Reservoir modeling

2

2.1 Introduction

Reservoir modeling means construction of a petroleum reservoir model for improving estimation of reserves, predicting reservoir performance, increasing production and making decisions regarding the development of the field. The main purpose of reservoir modeling is reservoir management. Reservoir modeling involves teamwork and people of different disciplines—geophysics, geology, petrophysics, mathematics, chemical and petroleum engineering—working together to construct a corresponding model.

A reservoir model consists of the following main parts: defining and specifying of objectives; data gathering and data analysis; building the reservoir model; conducting history-matching; conducting forecast scenarios; and reporting. The most important step in a successful reservoir study is to specify the objectives. The type and border of the model, quantity of data, quality of history-matching and forecast scenarios all depend on the objectives of the study. In specifying the objectives, the recovery mechanisms, quantity and quality of the available data and the schedule of the study have to be addressed properly.

One of the main factors in specifying the objectives of a study is to determine the recovery mechanisms of the reservoir, as reservoir modeling is only able to answer questions about the production history. Even if the reservoir model matches the field history of a reservoir with natural drive mechanisms, model results for secondary and tertiary processes are not reliable. This is because the mechanisms of primary (natural) recovery differ from the secondary and tertiary. For example, if there is an undersaturated petroleum reservoir with some impermeable shale layers and without an active aquifer, the primary recovery mechanism of the reservoir is fluid expansion and pore compaction. Therefore, the shale layers are not so important in fluid production from the reservoir. However, if after a period of fluid production water is injected into the reservoir (secondary recovery), the presence of impermeable shale layers is important. So, for a reliable forecast result from the reservoir model, a corresponding history is important. When defining the objectives of the study, quality and quantity of available data should be considered; for example, to have reliable modeling results of a gas injection scenario, reliable gas relative permeability data and capillary pressure data are essential.

Performance of hydrocarbon reservoirs is modeled by three methods: analogy, laboratory and mathematical methods. By the analogy method, reservoir characteristics of a similar reservoir are used rather than the original reservoir. This method is usually applied when the data of the original reservoir are meager. In laboratory methods, reservoir behavior is studied at laboratory scale and the results are then scaled up to the reservoir scale. Scaling up from laboratory to field is the most challenging issue when using laboratory methods. In mathematical methods, equations

Experimental Design in Petroleum Reservoir Studies. DOI: http://dx.doi.org/10.1016/B978-0-12-803070-7.00002-8

of mass conservation, energy conservation, fluid flow through porous media and Darcy's law are solved by numerical and analytical techniques. Odeh in 1982 listed the major steps used in mathematical modeling of a reservoir: formulation (writing equations of mass and/or energy balance, and utilizing empirical rules) to derive partial differential equations (PDEs), discretization of the PDEs to algebraic equations, linearization of the algebraic equations to linear equations, solution of the linear equations (by analytical or numerical techniques) to reach pressure and saturation distribution and well rates, and validation and application of the results.

2.2   Sources of data for reservoir modeling

Various types of data are used in reservoir modeling and multidisciplinary experts (geophysicists, geologists, petrophysicists, drilling engineers, reservoir engineers and process engineers) need to work in collecting the data. Some of these data (geological, geophysical and petrophysical data) are about the static state of the reservoir (frame, structure and well locations). Other data, such as pressure-volume-temperature (PVT) and production data, provide information on fluid movement in the reservoir and are considered to be dynamic data. Scales of the provided data are different, too; some data have point-scale (a few centimeters, such as petrophysical data) while other data are considered to be field source (such as seismic data, production and well-testing data). Fanchi (2010) defined four applicable scales in reservoir modeling: micro-scale (such as thin section and grain properties), macro-scale (such as rock and fluid properties), mega-scale (such as well logging data) and giga-scale (such as geophysical data).

Schulze-Riegert and Ghedan (2007) summarized different sources for reservoir modeling. They classified sources as sources of static data and sources of dynamic data. Static data sources are the sources to provide reservoir structure, thickness, fluid contacts, reservoir geometry, facies, grain size distribution, and pore compressibility. These sources are geophysical data, well logging data, core data, nuclear magnetic resonance (NMR), thin section analysis, special core analysis, well testing and facies analysis.

Dynamic data sources provide information about fluid composition, fluid PVT properties, fluids interfacial tension (IFT) data, fluid saturation, wettability, capillary pressure and relative permeability data. Dynamic data sources are core data, PVT experiments, well log data, well testing data and special core analysis.

2.3   Reservoir characterization

The first step in reservoir modeling is reservoir characterization, where required data are gathered, analyzed, and then a geologic (static) model is constructed. The static model is a fine grid model in which the geometry, distribution of petrophysical properties, and flow properties of a petroleum reservoir are, accurately and quantitatively, characterized. In fact, reservoir characterization is a basic step in reservoir modeling.

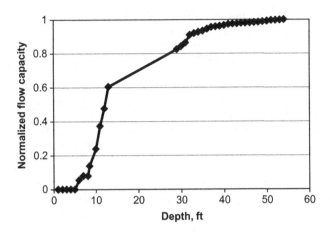

Figure 2.1 Identifying flow units using modified Lorenz plot.

Basic data in reservoir characterization are seismic data, well logging data, core data, well testing data, and rock and fluid data. In addition, all available data from drilled wells with interpretation of seismic data are used in reservoir modeling. Form, structure, shape, rock and fluid properties of the reservoir model should represent the original reservoir.

Since reservoirs are heterogeneous, in the characterization of rock and fluid throughout a reservoir, flow units must be considered. Each flow unit has a similar average of petrophysical properties (porosity, permeability, compressibility, fluid saturations), and has similar geologic properties (texture, mineralogy, sedimentary structure, bedding contacts, permeability barriers) [Lake, 1986; Fanchi, 2010]. A reservoir may contain several flow units. Identification of flow units could be done using a modified Lorenz plot by plotting normalized flow capacity as a function of depth. A change in slope of this plot is interpreted as a change in flow unit [Fanchi, 2010]. Figure 2.1 shows a typical example of identifying flow units.

2.3.1 Geophysical and geological data

Geological and geophysical data are essential elements to build a reservoir model structure; geophysical data help to identify reservoir borders and from geological data internal structure (skeleton) of the reservoir model is prepared. One of the most common types of geophysical data is seismic data. Seismic data are generated by producing shockwaves on the surface, transferring the waves into formations and measuring (recording) the reversing time of the waves to the surface. These recordings are then analyzed (processed) and interpreted by geophysicists.

Geophysicists use seismic surface sections to construct structural maps (top and bottom surfaces), fault detection, change in layer thickness, and layer continuity and pinchout. If the seismic survey is repeated over a producing hydrocarbon reservoir (time-lapse or 4-D seismic), the changes occurring in the reservoir (e.g., change in fluid saturation) could be determined by comparing the repeated datasets.

Geologists' activities in the reservoir characterization phase are usually summarized into four categories [Harris, 1975]:

- Rock studies: In rock studies, lithology and sedimentation media are investigated to determine the reservoir rocks.
- Reservoir structure studies: Here, the three-dimensional reservoir structure is built and continuity of reservoir properties as well as the rock thickness trends are investigated.
- Study on quality of reservoir properties: Here, variations of reservoir properties (such as porosity, permeability, and capillary) throughout the reservoir are studied.
- Integrated studies: Pore volume and fluids transmissibility are investigated.

In a constructed geological model, the petrophysical properties that are consistent with the observed data are estimated. Because of limited information and data, reservoir properties of unknown regions (unsampled locations) are estimated from known regions (sampled locations). In reality, the reservoir description involves the use of statistics (the technique is called geostatistics, a branch of statistical science that studies spatially distributed properties), where some types of implicit relationships between the data are assumed. The idea behind geostatistics is that underlying geological surfaces have continuity and hence they should be spatially related [Davis, 2002].

Variogram (or semivariogram) is the most commonly used method to describe the spatial relationship between geological properties. Suppose measured values of a geological property (for example, porosity) at two locations Δ intervals away are x_i and $x_{i+\Delta}$. Variogram $\gamma(\Delta)$ is defined as

$$\gamma(\Delta) = \frac{1}{2n} \sum_{i=1}^{n-\Delta} [x_i - x_{i+\Delta}]^2 \tag{2.1}$$

where n is the number of points.

The variogram increases as the distance (Δ) between the values increases. The variogram normally starts from zero at the origin and increases with the distance. Increase in the variogram is an indication of a weaker relationship between the data. A region where the variogram becomes flat is called the sill. The distance where the sill region starts is the range. The range is an indication of anisotropy in the geological property. The shorter is the range, the more anisotropy there is in the property. Usually, for geological properties, the range in the vertical direction is shorter than in the horizontal direction.

Let's consider an example of vertical variation of porosity in a well drilled 87 meters at the crest of a reservoir. The porosity was measured using a sonic log every 10 cm of the formation, Figure 2.2. The purpose is to generate a variogram for the first 2 meters of the formation. Figure 2.3 illustrates the generated variogram (using Eq. 2.1) of porosity in the first 2 meters of the formation.

When the variogram is plotted versus distance, useful information is discovered to help geologists and reservoir engineers in the reservoir characterization task. The behavior of the variogram at the origin and the range are the two important factors for geological continuity of the property and the degree of anisotropy in spatial continuity.

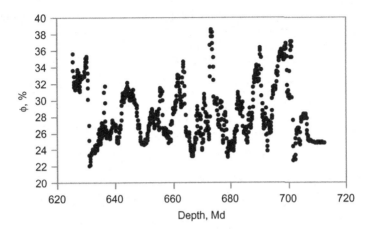

Figure 2.2 Variation of porosity versus depth.

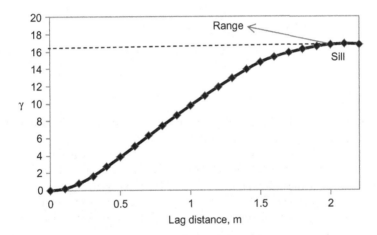

Figure 2.3 Variogram of porosity for the first 2 meters.

Geologists identify four main types of behavior of the variogram near the origin [Sarma, 2009]: parabolic behavior (a characteristic of a highly continuous property), linear behavior (an indication of a continuous property), nugget effect (a discontinuity at the origin indicating sampling error or fine-scale fluctuations of the property at short distances), and pure nugget effect (a flat line parallel to the X-axis representing a purely random property).

The variation of the ranges as a function of direction results in three possibilities: isotropy (the ranges are the same in all directions), geometrical anisotropy (the range changes with direction but by simple linear transformation of the coordinates it reduces to isotropic), and zonal anisotropy (indicating strongly directional variation of the property).

In order to use the variogram for estimating values at unsampled locations, variograms are modeled. Modeling the variogram is usually done by one of the four continuous functions of spherical, exponential, Gaussian and linear [Davis, 2002]. The equations of these models are as follows:

Spherical

$$\gamma(h) = (sill)\left[\frac{3h}{2(range)} - 0.5\left(\frac{h}{range}\right)^3\right] \tag{2.2}$$

Exponential

$$\gamma(h) = (sill)\left[1 - \exp\left(\frac{-3h}{range}\right)\right] \tag{2.3}$$

Gaussian

$$\gamma(h) = (sill)\left[1 - \exp\left(-\left(\frac{h}{range}\right)^2\right)\right] \tag{2.4}$$

Linear
$$\gamma(h) = (slope)h \tag{2.5}$$

In the spherical model, the variogram near the origin has linear behavior. The curve then rises to reach the sill and remains constant.

Like the spherical model, the exponential model behaves linearly near the origin. However, the curve cannot reach the sill at the range. It approaches the range asymptotically (i.e., where h approaches to positive infinity).

To model the parabolic behavior of the variogram at the origin, the Gaussian model is applied. By the Gaussian model, the geological property continuously and smoothly rises over short distances [Davis, 2002].

Figure 2.4 shows typical spherical, exponential and Gaussian modeling of the variogram.

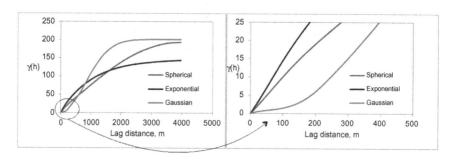

Figure 2.4 Typical spherical, exponential and Gaussian modeling of variogram.

The linear equation is the simplest model for variogram modeling. This model never reaches to sill and rises without limit.

To estimate the geological properties at unsampled locations, kriging technique (one of the most commonly known modeling techniques in geostatistics) is applied. Kriging is similar to regression technique in that to estimate a value at an unsampled location it uses values at the neighboring locations, weights (λ_i) assigned to the neighboring values and a linear relationship [Davis, 2002]:

$$X^*(u_0) = m\left(1 - \sum_{i=1}^{n} \lambda_i\right) + \sum_{i=1}^{n} \lambda_i X(u_i) \tag{2.6}$$

In Eq. 2.6 the observation variable at point location u_i is denoted by $X(u_i)$ and $X^*(u_0)$ is an estimate of the variable at unsampled location u_0.

In Eq. (2.6) the mean value, m, is assumed to be constant.

The difference between kriging and linear regression is that in kriging the variables are not independent.

Weights (λ_i) are estimated using the following constraint:

$$\sum_{i=1}^{n} \lambda_i \text{cov}(u_i, u_j) = \text{cov}(u_0, u_j) \tag{2.7}$$

where $\text{cov}(X(u_i), X(u_j))$ is the covariance value between points located at u_i and u_j and $\text{cov}(X(u_0), X(u_j))$ is the covariance value between unsampled location u_0 and sampled location u_j.

Covariance is a measure of linear correlation between two random variables. For two random variables of $X(x_i)$ and $Y(y_i)$ it is calculated as:

$$\text{cov}(X, Y) = \frac{n \sum_{i=1}^{n} x_i \cdot y_i - \sum_{i=1}^{n} x_i \sum_{i=1}^{n} y_i}{n(n-1)} \tag{2.8}$$

Equations (2.7) and (2.8) depict that a set of simultaneous equations should be solved to find the weights λ_i.

Once the value at the unsampled location is estimated, the standard error of the estimated value is found using the following equation [Davis, 2002]:

$$\sigma = \sqrt{\sum_{i=1}^{n} \lambda_i \text{cov}(u_0, u_i)} \tag{2.9}$$

In kriging, uniformly distributed data throughout the study area are required to get good estimates.

Kriging is better illustrated by a simple example. Suppose the top of a reservoir is measured in three drilled wells (W_1, W_2, W_3) and it is desired to estimate the top at two points P_1 and P_2, Figure 2.5.

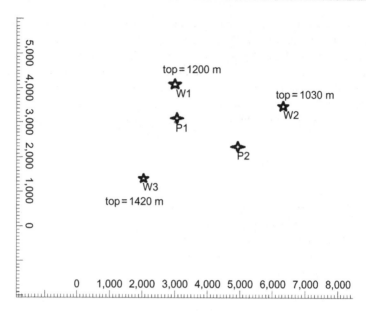

Figure 2.5 Map showing coordinates of wells and reservoir top at three drilled wells.

For location P1, we should solve Eq. 2.7 to find the weights:

$$\lambda_1 cov(W_1, W_1) + \lambda_2 cov(W_2, W_1) + \lambda_3 cov(W_3, W_1) = cov(P_1, W_1)$$
$$\lambda_1 cov(W_1, W_2) + \lambda_2 cov(W_2, W_2) + \lambda_3 cov(W_3, W_2) = cov(P_1, W_2)$$
$$\lambda_1 cov(W_1, W_3) + \lambda_2 cov(W_2, W_3) + \lambda_3 cov(W_3, W_3) = cov(P_1, W_3)$$

Calculating the covariance using Eq. 2.8 yields a system of three equations, three unknowns:

$$13.42\lambda_2 + 11.52\lambda_3 = 4$$
$$13.42\lambda_1 + 19.14\lambda_3 = 13.3$$
$$11.52\lambda_1 + 19.14\lambda_2 = 7.89$$

By solving this system, weights are found as

$$\lambda_1 = 0.5904; \quad \lambda_2 = 0.0569; \quad \lambda_3 = 0.2809$$

Now, Eq. 2.6 is used for estimating the top of reservoir at location P1:

$$P1 = \left(\frac{1200 + 1030 + 1420}{3}\right)(1 - 0.9282) + 0.5904(1200)$$

$$+ 0.0569(1030) + 0.2809(1420)$$

$$P1 = 1253.32$$

The standard error using Eq. 2.9 is:

$$\sigma = \sqrt{(0.5904)(4) + (0.0569)(13.297) + (0.2809)(7.889)}$$
$$= \sqrt{5.33} = 2.3\,\text{m}$$

When this procedure is repeated for location P2, the weights are found as:

$$\lambda_1 = 0.1538; \quad \lambda_2 = 0.5417; \quad \lambda_3 = 0.2778$$

and the reservoir top is estimated to be 1169.3 m. The standard error in this case is 3 m.

In kriging, squared error (difference between the estimated value and the true value at the unsampled location) is locally minimized to estimate a best value (a value that is as close as possible to the true value) at the unsampled location. However, in order to model reservoir behavior as accurately as possible, the errors should be minimized globally. If global minimization of errors (existence of overall geological continuity and reservoir heterogeneity interpreted from all available reservoir engineering information and data) is provided during the property estimation process, the unsampled value is estimated independently of any estimation at other unsampled locations [Caers, 2005].

Therefore, in addition to the kriging that is an estimation method for smooth interpolation (or extrapolation) among data, stochastic simulation techniques are implemented to regenerate what is actually seen in the data (geological continuity and reservoir heterogeneity) as well as to match values at sampled locations. In contrast to estimation methods, simulation methods can result in multiple outcomes.

Kelkar and Perez [2004] classified the simulation techniques into grid-based and object-based simulation methods. In grid-based simulation methods, the reservoir is discretized into several homogeneous grid-blocks with individual properties. Object-based simulation techniques generate values based on the shape and size of geological objects. Objects could be two-dimensional or three-dimensional [Kelkar and Perez, 2004]. Among grid-based techniques, Gaussian methods are the most commonly used techniques.

An example of application of Gaussian grid-based simulation is illustrated here. In this example, there are eight wells (W-1, W-2, W-3, W-4, W-5, W-6, W-7, W-8) drilled in a reservoir with a gross bulk volume of $1.5 \times 10^9\,\text{m}^3$ (6800 m $\times$ 4700 m $\times$ 47 m). Two geological units within three markers are recognized by geologists. The correlation between the well tops is shown in Figure 2.6. By using inverse distance weighting estimation, the surface of each geological formation is created. Inverse distance weighting is a method for interpolation which is based on nearby data values weighted by distance [Shepard, 1968]:

$$u = \frac{\sum\left(\dfrac{u_i}{d_i^{\alpha}}\right)}{\sum \dfrac{1}{d_i^{\alpha}}} \tag{2.10}$$

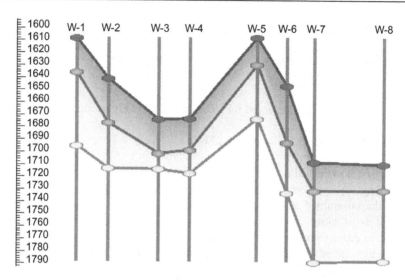

Figure 2.6 Top and bottom of the reservoir formations for the eight wells.

where u_i is the value at a nearby location, d_i is the distance between known and unknown points and α is the exponent value.

The result with an exponent value of 2 is shown in Figure 2.7.

The reservoir is then discretized to 136 divisions in the x direction, 94 divisions in the y direction, and 12 divisions in the z direction to create 153,408 ($136 \times 94 \times 12$) grid-blocks (Figure 2.8). Each grid cell has 50 m length and 50 m width.

Four wells (W-1, W-2, W-3, W-4) out of eight wells have porosity logs (Figure 2.9) and they are used to generate porosity for the whole reservoir. Horizontal variation of measured porosity by sonic log in each layer is illustrated in Figure 2.10. Using Gaussian grid-based simulation techniques and variogram modeling, the porosities for whole grids are then generated. For comparison, three different variogram models (exponential, Gaussian and spherical) are used and the generated porosities are compared to each other. Generated porosities are shown in Figures 2.11 to 2.13. A summary of using different variogram models for porosity estimation in the reservoir is shown in Table 2.1. The table depicts that, by exponential modeling, the porosities of about 72% of cells are in the range of 0.16 to 0.27. Therefore, most of the oil in place is accumulated in the "porous" cells. If one applies Gaussian models, then about 30% of cells have porosities of 0.16 to 0.27 and the porosities of 25% of the cells are in the range of 0.35 to 0.39. One-third of cells have porosities in the range of 0.01 to 0.05. The spherical model results in 58% of cells with porosities of 0.16 to 0.27. In the spherical model, like the exponential model, about 2% of cells have porosities greater than 0.27. The spherical model, similar to the Gaussian model, generates porosities of less than 0.16 for about 40% of cells. About 20% of cells in

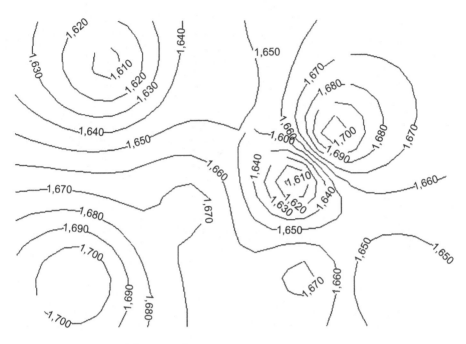

Figure 2.7 Contours of the reservoir top.

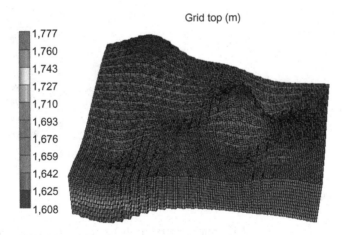

Figure 2.8 Discretization of the reservoir to 153,408 grid-cells.

the spherical model have porosities of 0.01 to 0.05. This value for the exponential model is less than 10%. The generated data show that the average estimated porosities by applying exponential and Gaussian models are very close to each other (0.191 and 0.189) and both are greater than the value generated by the

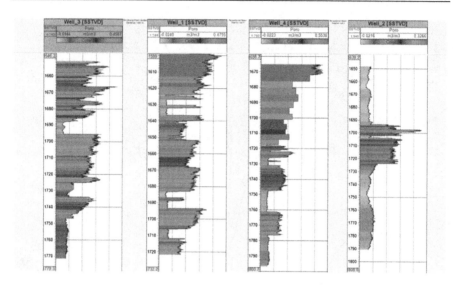

Figure 2.9 Porosity logs of four wells drilled in the reservoir.

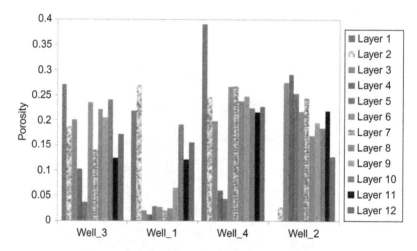

Figure 2.10 Horizontal variation of porosity.

spherical model (0.159). Different reservoir behaviors are expected from different generated models. But which result should be used for the actual reservoir study? This example depicts that geostatistical simulation results should be considered for uncertainty analysis.

In one more attempt, the porosities are generated by using an estimation method of kriging. For comparison, generated porosities of layer 2 by kriging and the Gaussian simulation technique are shown in Figure 2.14. Since there is no measured data in wells 5 to 8, kriging smoothly interpolates (or extrapolates) between the

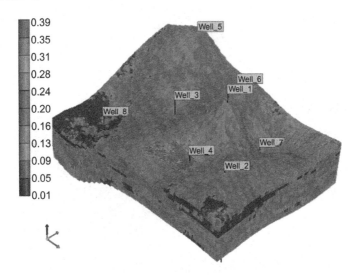

Figure 2.11 Estimated porosities by using exponential modeling of variogram.

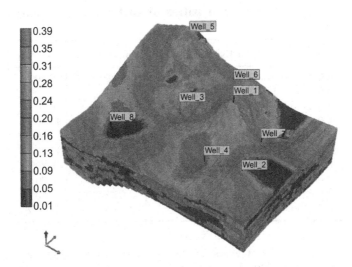

Figure 2.12 Estimated porosities by using Gaussian modeling of variogram.

data (wells 1 to 4) to generate a best estimation of porosities at unsampled locations (left picture). However, in the Gaussian simulation technique, in addition to matching the measured porosities at wells 1 to 4, it attempts to generate a model representing geological continuity in terms of porosity (right picture).

In the geostatistical generation of geological property distributions, in addition to errors (or uncertainties) due to using different variogram models and/or estimation methods, the nugget effect (caused by measurement error and/or fine-scale

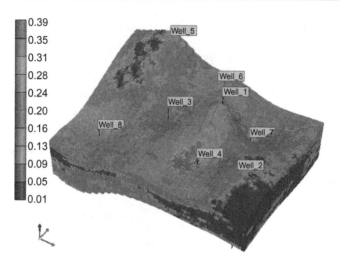

Figure 2.13 Estimated porosities by using spherical modeling of variogram.

Table 2.1 Generated porosities by different variogram modeling

Interval	Exponential	Gaussian	Spherical
From 0.35 to 0.39	0.50%	25.20%	0.10%
From 0.31 to 0.35	0.40%	0.80%	0.20%
From 0.27 to 0.31	1.60%	1.70%	0.80%
From 0.23 to 0.27	24.30%	10.40%	15.50%
From 0.20 to 0.23	29.20%	10.90%	26.30%
From 0.16 to 0.20	19.20%	8.50%	17.20%
From 0.12 to 0.16	6.80%	3.90%	6.80%
From 0.08 to 0.12	5.40%	3.40%	6.00%
From 0.05 to 0.08	3.70%	2.60%	4.80%
From 0.01 to 0.05	9.00%	32.60%	22.30%

fluctuations of the property) would propagate errors in the reservoir modeling. Pawar and Tartakovsky [2000] examined propagation of errors through reservoir modeling where permeability distributions were generated using a variogram model. They applied a power-law variogram model with nugget effects (nugget 3 and nugget 7) based on the measurement errors. The generated permeability distributions were then used for simulating the fluid flow through a reservoir for 1600 days. The results were compared to the simulation results obtained from a "ground truth" model (a reservoir model with true permeability distribution). They repeated the examination for a case with nugget zero. The study showed significant differences in the oil production when compared to the true model. The difference increased with the nugget value [Pawar and Tartakovsky, 2000].

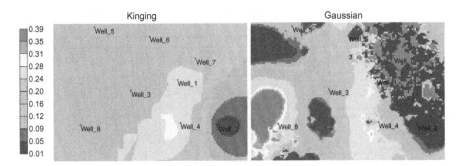

Figure 2.14 Estimated porosities of layer 2 by kriging and Gaussian simulation techniques.

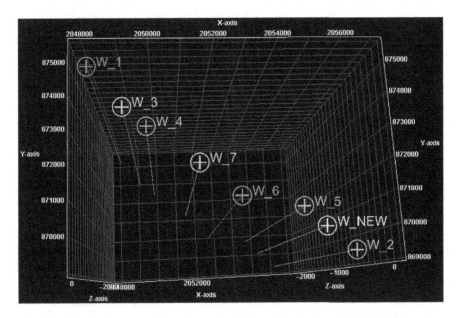

Figure 2.15 Well locations in the carbonate reservoir.

As one more example of error propagation during geostatistical generation of geological properties, a case of porosity distribution through a carbonate reservoir is elucidated in the following. In a carbonate reservoir with seven drilled wells (W_1 to W_7), porosity distribution was generated using a Gaussian simulation technique. To do this, a variogram of porosity data of the seven wells was matched by an exponential model with nugget 0.167, sill 1.05 and range 63. Two years after constructing the geologic model, a new well (W_NEW) was drilled in the reservoir and a full set of petrophysical logs was run inside the well. Well locations are shown in Figure 2.15. Now, we would like to compare the porosity of the new well with what we had already generated by geostatistical techniques. For better comparison, the porosity log is resampled with constant interval steps, using linear interpolation, Figure 2.16.

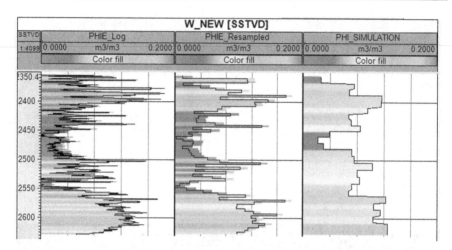

Figure 2.16 Log porosity versus simulator porosity.

The comparison shows a qualitatively good agreement between the log data and the simulator values. As is mentioned, this is a positive point of the Gaussian simulation technique. However, if one quantitatively compares the porosities, he would discover 18% difference. This error, surely, would affect volume of hydrocarbon in-place as well as the reservoir behavior and volume of hydrocarbon production.

2.3.2 Engineering data

While geological data are about rock properties and the predominant processes (diagenesis and sedimentation) on rock property distributions, engineering data correspond to static and dynamic behavior of fluids in the reservoirs. Most sources of engineering data and geological data are the same; however, analyzing methods of the data are different.

2.3.2.1 Core data

Core laboratories explain the behavior of fluids and rock under controlled conditions. In the laboratory, core appearance, routine and special core tests are analyzed and the results are then applied for reservoir dynamic modeling.

Core appearance represents reservoir lithology, presence of shale layers, gross thickness and net-to-gross ratio of the reservoir. In routine core tests major properties of rock such as porosity, absolute permeability and initial fluid saturations are measured. More complicated rock properties such as compressibility, relative permeability, capillary pressure and end-point saturations are measured in special core tests.

It should be noted that the results of core laboratories are usually scaled up from laboratory scale to reservoir scale. This step is called upscaling and is one of the most challenging issues of heterogeneous reservoirs. Generally, there are two types of averaging and upscaling in reservoir modeling. In the first type, sizes of geological grids (fine grids) are increased (coarse grids). In the second, results of core laboratory tests conducted on core of a few centimeters size are upscaled to coarse grids. Averaging and upscaling raise uncertainty of reservoir modeling and so should be used as infrequently as possible.

Averaging methods are classified into power averaging (such as arithmetic mean $\left(x_a = \frac{\sum x_i}{n}\right)$, geometric mean $\left(x_g = \sqrt[N]{\prod^n x_i}\right)$ and harmonic mean $\left(x_h = \frac{n}{\sum \frac{1}{x_i}}\right)$), renormalization techniques, the method of pressure equation solution, the tensor method and pseudo-functions technique.

For additive properties, arithmetic averaging normally gives reliable results. Thus, for gross thickness (h), porosity (ϕ), and water saturation (S_w) we can use the following equations [Ertekin et al., 2001]:

$$\bar{h} = \frac{\sum A_i h_i}{\sum A_i}$$

$$\bar{\varphi} = \frac{\sum \varphi_i h_i}{\sum h_i} \quad (2.11)$$

$$\bar{S} = \frac{\sum S_{w_i} \varphi_i h_i}{\sum \varphi h_i}$$

Upscaling of effective permeability is more challenging than other properties [King, 1996]. Arithmetic and harmonic averaging are the most commonly used methods. When flow occurs along parallel layers of constant permeability, arithmetic average (k_a) provides the effective permeability of the upscaled layer (coarsened). In the case of series layers which are perpendicular to flow direction, the harmonic average (k_H) is used:

$$\bar{k}_a = \frac{\sum k_i h_i}{\sum h_i} \quad (2.12)$$

$$\bar{k}_H = \frac{\sum h_i}{\sum \frac{h_i}{k_i}} \quad (2.13)$$

2.3.2.2 Well logging data

Well logging could determine formation properties at reservoir scale. These properties are used by geologists, reservoir engineers and petrophysicists. Some formation properties such as porosity, initial fluid saturations, lithology, net thickness and pressure gradient of fluids are directly measured by well logging. Figure 2.17 shows examples of gamma ray, density, water saturation and porosity logs. Some other

(a)

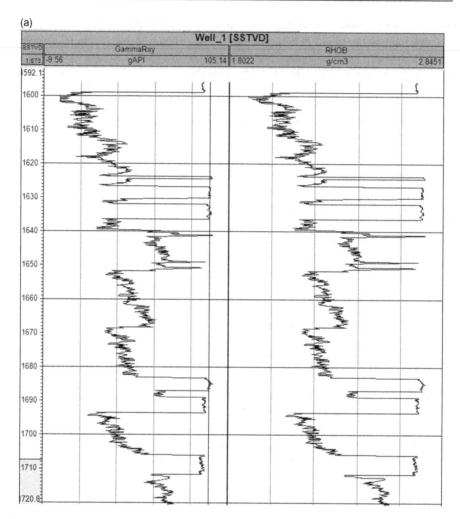

Figure 2.17 (a) Examples of gamma ray and density logs.

properties are derived from directly measured properties; for example, having porosity, the reservoir permeability is estimated; see Figure 2.18.

2.3.2.3 Pressure transient data

This type of data, like well logging data, gives some useful information at reservoir scale. Analysis of pressure transient data prepares information such as horizontal permeability, skin, well geometric factor, buildup pressure and average static pressure of the reservoir. Effective horizontal permeability (relative to oil or gas) calculated from pressure transient analysis is one of the most important sources for reservoir permeability. Pressure transient data are provided by analytically solving the diffusivity equation and using well testing data (pressure buildup or drawdown test).

(b)

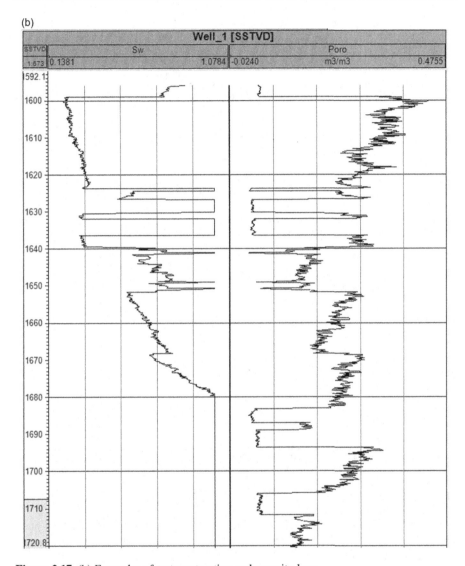

Figure 2.17 (b) Examples of water saturation and porosity logs.

The diffusivity equation of radial flow of slightly compressible fluids in a homogeneous horizontal reservoir may be written as [Ahmed & Meehan, 2011]:

$$\frac{\partial^2 p}{\partial r^2} + \frac{1}{r}\frac{\partial p}{\partial r} = \frac{1}{\eta}\frac{\partial p}{\partial t}$$

$$\eta = \frac{c \cdot k}{\varphi \mu c_t}$$

(2.14)

where c is a conversion factor.

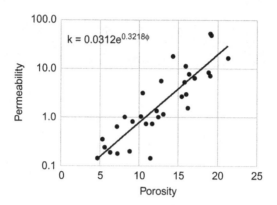

Figure 2.18 Typical permeability-porosity semilog plot.

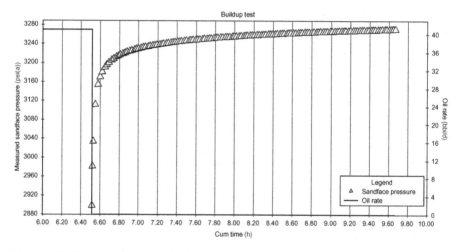

Figure 2.19 Example of pressure buildup test.

Figure 2.19 shows an example of a buildup test; a well that produced oil at a rate of 41 stb/day for 6.53 hours was shut in and its bottom-hole pressure measured. The test lasted for 3 hours.

The data were then analyzed to estimate permeability and initial pressure of the reservoir. Analysis is shown in Figure 2.20.

After analyzing the data, it is possible to model the test to estimate reservoir volume and original oil-in-place, Figure 2.21.

2.3.2.4 Properties of reservoir fluids

Basic knowledge of the pressure-volume-temperature (PVT) relationships of oil, gas and water phases is needed when studying a petroleum reservoir. Depending on reservoir temperature and pressure, hydrocarbon reservoirs are divided into five

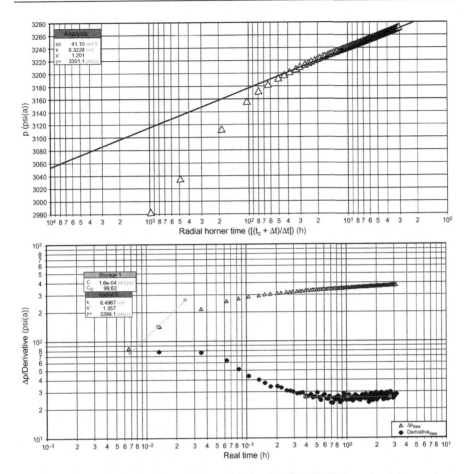

Figure 2.20 Analysis of pressure buildup test (using F.A.S.T. WellTest software).

categories: black oil reservoirs, volatile oil reservoirs, gas condensate reservoirs, wet gas reservoirs, and dry gas reservoirs (Figure 2.22).

In black oil reservoirs, temperature and pressure of reservoir fluid are far from the critical point. The fluid in the reservoir originally exists as a single-phase liquid and becomes two phases when pressure is below the bubble point pressure. The liquid volume of single-phase black oil increases as the reservoir pressure decreases and it shrinks where pressure drops below the bubble point. Figure 2.23 shows a typical phase envelope for black-oil fluids.

Volatile reservoirs have temperature and pressure somewhat nearer the critical point. These oils are characterized by rapid shrinkage below the bubble point. Figure 2.24 shows a typical phase envelope for volatile oils.

A reservoir fluid in which the temperature and pressure are located to the right of the critical point is known as a gas condensate reservoir. The fluid in the reservoir originally exists as a single-phase gas and becomes two phases as pressure

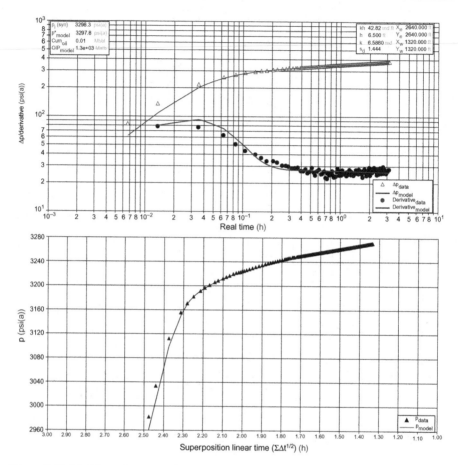

Figure 2.21 Modeling of pressure buildup test to estimate reservoir volume and original oil in place (using F.A.S.T. WellTest software).

drops below dew point pressure. Below the dew point pressure, both gas and liquid (retrogrades) fluids are produced and the fluid remaining in the reservoir has variable composition but constant reservoir volume. Figure 2.25 shows a typical phase envelope for gas condensate fluids.

To predict variations in reservoir pressure and other flow parameters, full PVT properties have to be determined and defined in the reservoir model. For that purpose several experiments are conducted on the sampled fluids in the PVT laboratory. These experiments are done in two phases: preliminary and full experiments. In preliminary experiments, specific gravity, gas oil ratio (GOR) and saturation pressure are measured and reported.

Here are brief definitions of important PVT properties:

Specific Gravities: Specific gravity is defined as a ratio of the density of a substance at well-defined conditions to the density of a known material such as gas or

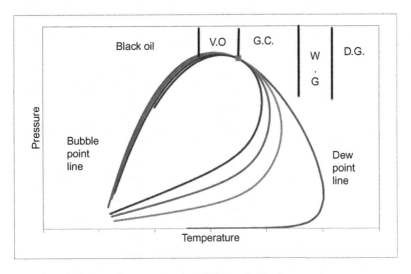

Figure 2.22 Typical phase diagram showing different fluids of reservoirs.

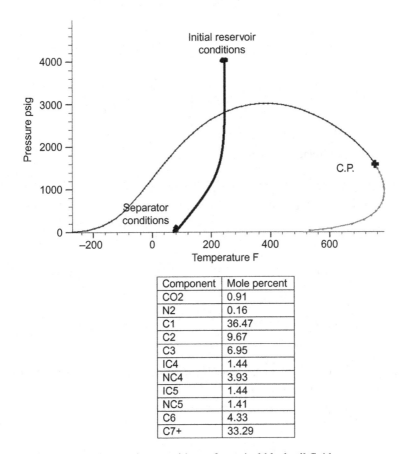

Component	Mole percent
CO2	0.91
N2	0.16
C1	36.47
C2	9.67
C3	6.95
IC4	1.44
NC4	3.93
IC5	1.44
NC5	1.41
C6	4.33
C7+	33.29

Figure 2.23 Phase envelope and compositions of a typical black-oil fluid.

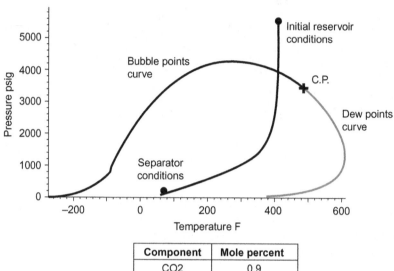

Component	Mole percent
CO2	0.9
N2	0.3
C1	53.47
C2	11.46
C3	8.79
C4	4.56
C5	2.09
C6	1.51
C7+	16.92

Figure 2.24 Phase envelope and compositions of a typical volatile oil.

water at similar standard conditions. For a gas system the specific gas gravity is defined as:

$$\gamma_{gas} = \frac{\rho_{gas}(T_{sc}, P_{sc})}{\rho_{air}(T_{sc}, P_{sc})} \qquad (2.15)$$

For a liquid system:

$$\gamma_{oil} = \frac{\rho_{oil}(T_{sc}, P_{sc})}{\rho_{water}(T_{sc}, P_{sc})} \qquad (2.16)$$

The gravity widely used in the petroleum industry is the American Petroleum Index or API Gravity (°API) defined as:

$$°API = \frac{141.5}{\left(\frac{60}{60}\gamma_{oil}\right)} - 131.5 \qquad (2.17)$$

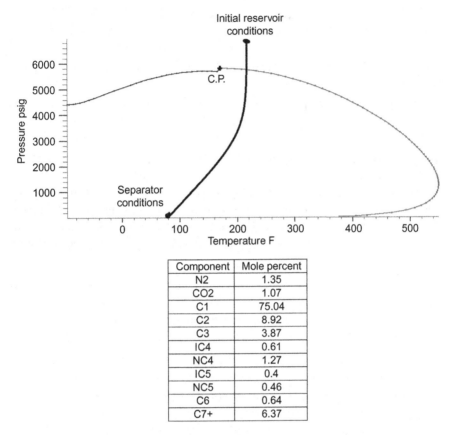

Figure 2.25 Phase envelope and compositions of a typical gas condensate fluid.

Isothermal Compressibility: The Isothermal Compressibility (C) of a fluid is defined as the relative change in volume ($\Delta V/V$) per unit change in pressure (P):

$$C = -\left(\frac{1}{V}\right)\left(\frac{\partial V}{\partial P}\right)_T \tag{2.18}$$

For oil compressibility the above equation is rearranged to give:

$$C_o = -\left(\frac{1}{V}\right)\left(\frac{\partial V}{\partial P}\right)_T = -\left(\frac{\partial(\ln(V))}{\partial P}\right)_T \tag{2.19}$$

After integration:

$$V = V_i e^{-C_o(P-P_i)} \tag{2.20}$$

A similar equation can be written using density:

$$\rho = \rho_i e^{C_o(P-P_i)} \tag{2.21}$$

As C_o is small for liquids (and $e^x \approx 1 + x$, where x is small), the above equations are often written as:

$$V = V_i[1 - C_o(P - P_i)] \tag{2.22}$$

$$\rho = \rho_i[1 - C_o(P - P_i)] \tag{2.23}$$

where the subscript i refers to the initial (or reference) conditions and C_o is assumed to be constant.

Formation Volume Factors: The change in volume of a reservoir fluid undergoing production is normally expressed in terms of the formation volume factor. In general terms, the formation volume factor (B) can be regarded as the ratio of volume of reservoir fluid at reservoir conditions (V_{res}) to volume of fluid at standard conditions (V_{sc}), i.e.,

$$B = \frac{V_{res}}{V_{sc}} \tag{2.24}$$

For an oil reservoir, the formation volume factor is the volume of reservoir oil per barrel of stock tank oil.

$$B_o = \frac{V_{res}}{V_{stb}} \tag{2.25}$$

For a dry gas reservoir it is the volume of reservoir gas per standard volume of surface gas (usually measured at 14.7 psia, 60°F).

$$B_g = \frac{V_{res}}{V_{sc}} \tag{2.26}$$

For gas, one can write:

$$B_g = \frac{V_{res}}{V_{sc}} = \frac{\left(\dfrac{Z_{res}nRT_{res}}{P_{res}}\right)}{\left(\dfrac{Z_{sc}nRT_{sc}}{R_{sc}}\right)} = \frac{Z_{res}T_{res}P_{sc}}{P_{res}Z_{sc}T_{sc}} \tag{2.27a}$$

or

$$B_g = \frac{Z_{res}T_{res}(14.7psia)}{(1.0)(60 + 460°R)P_{res}} = \frac{0.02827Z_{res}T_{res}}{P_{res}} \tag{2.27b}$$

Solution Gas-Oil Ratio: Solution gas-oil ratio, R_s, is defined as the volume of gas measured at standard conditions that will dissolve in one barrel of stock tank oil when subjected to reservoir pressure and temperature. The gas solubility is also commonly referred to as the gas to oil ratio (GOR). Rather than determine the gas that will dissolve in a certain amount of oil (R_s), it is customary to measure the gas released from reservoir oil (GOR) as the pressure decreases.

In a full PVT experiment, corresponding to fluid type, constant composition expansion (CCE), differential liberation and/or constant volume depletion tests are conducted to specify the following items:

- Fluid composition
- Saturation pressures
- Isothermal compressibility factor
- Gas compressibility factor
- Change in fluids formation volume factor with pressure
- Change in gas oil ratio with pressure
- Liberated gas composition at each pressure step
- Change in fluid density with pressure
- Density of residual oil
- Change in fluid viscosity with pressure

Constant Composition Expansion: The constant composition expansion (CCE) or constant mass expansion (CME) experiment is carried out in all PVT studies regardless of fluid type. It is used to measure the total fluid volume and compressibility over a wide range of pressures extending beyond initial reservoir pressure to pressures below the separator pressures. For black oils and volatile oils it is also used to determine the saturation pressure at reservoir conditions.

Differential Liberation: The differential liberation experiment (Diff. Lib.) is the classical depletion experiment for reservoir oils carried out at reservoir temperature to simulate the volumetric and compositional changes in the reservoir during production (depletion). Oil and gas formation volume factors, oil and gas densities, solution GOR (R_s), and the gas deviation factor (Z) as a function of pressure, and the gas expansion factor are the quantities that are determined from differential liberation experiments.

Constant Volume Depletion (CVD) Experiment: The constant volume depletion (CVD) experiment is conducted for volatile oil and gas condensate reservoir fluids to model the reservoir fluid depletion during production. Data include the volume of gas and liquid at each pressure stage, the cumulative produced (wet) gas, the gas Z-factor and the composition of the produced fluids. All the reported volumetric data are reported relative to the volume of the fluid at the dew point for gas condensate or bubble point for volatile oils.

Figure 2.26 shows schematic diagrams for CCE and differential liberation (Diff. Lib.) tests of a typical black oil.

Separator Tests: In the separator test, a known volume of the reservoir oil at its bubble point is flashed generally in two stages, where the last stage represents the stock tank. Separator tests are carried out on reservoir fluids to provide volumetric and other information on the stock tank oil and liberated gas streams.

Oil and Gas Viscosities: In PVT laboratory tests, oil viscosities at reservoir temperature and various pressures are usually measured by rolling ball viscometer (RBV). The gas viscosity is correlated accurately by using empirical correlations. There are several correlations for estimating oil and gas viscosities. Among them, the Lorentz-Bray-Clark correlation (1964), Jossi-Stiel-Thodos correlation, and Pederson correlation (1989) are the most commonly used correlations.

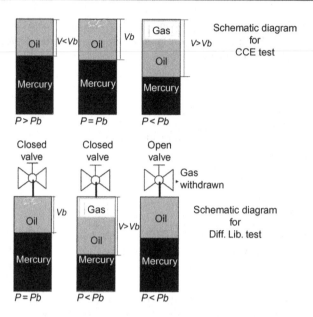

Figure 2.26 Schematic diagrams for CCE and Diff. Lib. Tests.

Interfacial Tension: Interfacial tension (σ) between gas and liquid phases of reservoir fluids is estimated using the Parachor correlation [Danesh, 1998]:

$$\sigma^{0.25} = (Parachor\_Factor)(\rho_L - \rho_V) \tag{2.28}$$

where indices L and V refer to liquid and gas phases, respectively.

Table 2.2 shows typical reported PVT data.

Generally, crude oil contains some solution gas, and some connate water always accompanies it. In most cases, it is acceptable to assume the compositions of oil and gas are constant and solubility of gas in oil is merely a function of pressure. Therefore, one can consider oil as a single-component material (or, in a more accurate expression, a single pseudo-component material). This is also applicable for gas. In cases where gas and oil compositions vary with location, the variations have to be considered. These variations are formulated using equations of state (EOS). One of the most commonly used EOS in the petroleum industry is Peng-Robinson [Peng and Robinson, 1976]:

$$
\begin{aligned}
p &= \frac{RT}{V - b} - \frac{a\alpha}{V^2 + 2bV - b^2} \\
a &= 0.457235 \frac{R^2 T_C^2}{p_C} = \Omega_a \frac{R^2 T_C^2}{p_C} \\
b &= 0.077796 \frac{RT_C}{p_C} = \Omega_b \frac{RT_C}{p_C} \\
\alpha &= (1 + \kappa(1 - T_r^{0.5}))^2 \\
\kappa &= 0.37464 + 1.54226\omega - 0.26992\omega^2
\end{aligned}
\tag{2.29}
$$

where c is for critical states and ω is the Pitzer acentric factor.

Table 2.2 Typical reported PVT data for black-oil modeling

GOR (sm3/ sm3)	Pressure (bar)	Bo (rm3/ sm3)	Oil viscosity (cP)	Pressure (bar)	Bg (rm3/ sm3)	Gas viscosity (cP)
11.46	40	1.064	4.338	40	0.02908	0.0088
17.89	60	1.078	3.878	60	0.01886	0.0092
24.32	80	1.092	3.467	80	0.01387	0.0096
30.76	100	1.106	3.1	100	0.01093	0.01
37.19	120	1.12	2.771	120	0.00899	0.0104
43.62	140	1.134	2.478	140	0.00763	0.0109
46.84	150	1.141	2.343	150	0.00709	0.0111
50.05	160	1.148	2.215	160	0.00662	0.0114
56.49	180	1.162	1.981	170	0.0062	0.0116
59.7	190	1.169	1.873	180	0.00583	0.0119
62.92	200	1.176	1.771	190	0.00551	0.0121
66.13	210	1.183	1.674	200	0.00521	0.0124
69.35	220	1.19	1.583	210	0.00495	0.0126
72.57	230	1.197	1.497	220	0.00471	0.0129
74	234.46	1.2	1.46	230	0.00449	0.0132
80	245	1.22	1.4	234.46	0.0044	0.0133

It should be noted equations of state were initially developed for pure compounds and, therefore, when used for mixtures they should be modified. This modification is accomplished by changing the parameters of EOS (in the Peng-Robinson EOS, a, b and ω change so that results of the EOS are matched to experimental results). This is done by regression techniques. The objective of the regression is to minimize the difference between experimental data and predicted results of the equation of state:

$$F = \sum w_i \left(\frac{y_{EOS} - y_{Exp}}{y_{Exp}} \right)^2 \tag{2.30}$$

where w_i is a weight factor and indices EOS and Exp refer to the equation of state and experiment, respectively.

Figure 2.27 is an example of tuning the Peng-Robinson equation of state for a constant volume depletion test. In this example, tuning the EOS is conducted by WinProp, a phase-behavior and fluid property package.

2.3.2.5 Rock-Fluid data

In reservoir modeling, the behavior of fluids in the vicinity of rock pores should be characterized. Relative permeability (k_r) and capillary pressure (P_c) data are the two terms used for rock-fluid characterization. These terms are then used for predicting the capacity of a reservoir in flowing of oil, water, and gas throughout the life of the reservoir.

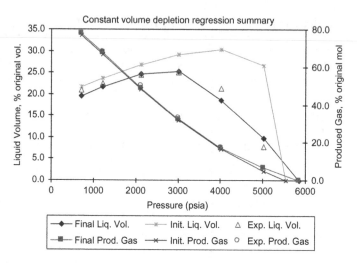

Figure 2.27 Tuning Peng-Robinson equation of state.

In multiphase flow through water-wet porous media (where water tends to contact most part of the porous media), capillary pressure (the difference between non-wetting phase pressure and wetting phase pressure) is the required force to push a hydrocarbon droplet through pores (works against the interfacial tension between nonwetting and wetting phases) [Bear, 1972]. In reservoir modeling, capillary pressure is required to determine initial water saturation of grid-cells. It is also needed to perform multiphase displacement (e.g. water injection and gas injection) calculations. Relative permeability of a phase is a dimensionless parameter and is defined as the ratio of effective permeability of that phase to a base permeability. It is required to model multiphase flow through reservoir. Relative permeability and capillary pressure strongly depend on the type of the fluids and rocks in a producing reservoir [Christiansen, 2001].

Most often, these two terms are measured in core laboratories. Three commonly used techniques for measuring capillary pressure data are: the porous plate technique, the centrifugal technique and the mercury injection technique.

Two major laboratory methods to measure relative permeability are steady-state and non−steady-state techniques.

Figures 2.28 and 2.29 show typical laboratory measured relative permeability and capillary pressure data.

In the case of no measured data, there are some empirical correlations that correlate relative permeability and capillary pressures to the saturation of phases in the porous media:

$$\text{krw} = \text{krw}_{(\text{Sorw})} \left(\frac{\text{Sw} - \text{Swc}}{1.0 - \text{Swc} - \text{Sorw}} \right)^{\text{nw}} \tag{2.31}$$

$$\text{krow} = \text{kro}_{(\text{Swc})} \left(\frac{\text{So} - \text{Sorw}}{1.0 - \text{Swc} - \text{Sorw}} \right)^{\text{no}} \tag{2.32}$$

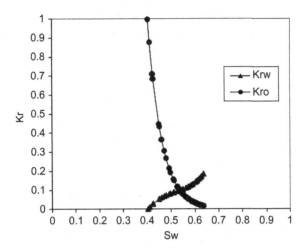

Figure 2.28 Laboratory measured water-oil relative permeability in a water-wet rock.

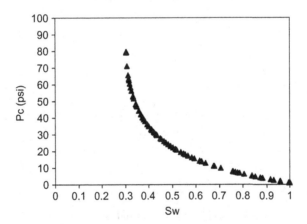

Figure 2.29 Laboratory measured oil-water capillary pressure curve.

$$\text{krog} = \text{kro}_{(Sgc)} \left(\frac{Sw - Swc - Sorg}{1 - Swc - Sorg} \right)^{nog} \tag{2.33}$$

$$\text{krg} = \text{krg}_{(Sorg)} \left(\frac{1 - Sw - Sgc}{1 - Swc - Sgc - Sorg} \right)^{ng} \tag{2.34}$$

And for capillary pressure:

$$P_{cow} = a_0 + a_1(1 - Sw) + a_2(1 - Sw)^2 + a_3(1 - Sw)^3 \tag{2.35}$$

$$P_{cgo} = b_0 + b_1 S_g + b_2 S_g^2 + b_3 S_g^3 \tag{2.36}$$

where coefficients of a_i and b_i are determined empirically.

It should be noted that, in the core laboratory, capillary pressure data are measured at laboratory conditions. The data then should be converted to reservoir conditions by using the Young-Laplace equation:

$$Pc = \frac{2\sigma Cos\,\theta}{r} \tag{2.37}$$

and

$$\frac{Pc_{res.cond.}}{Pc_{lab.cond.}} = \frac{\sigma_{res.cond.}Cos\,\theta_{res.cond.}}{\sigma_{lab.cond}Cos\,\theta_{lab.cond.}} \tag{2.38}$$

As stated, because reservoir heterogeneity, relative permeability and capillary pressure vary from place to place in a reservoir, this means that more than one rock type should be considered in the reservoir modeling; rock types with similar characteristics are classified into the same category.

Characterization of rock types is normally done using the normalized Leverett J-function (J^*) and normalized water saturation (Swn):

$$J^* = \frac{J\sigma Cos\,\theta}{Pc}\sqrt{\frac{\varphi}{k}} \tag{2.39}$$

$$Swn = \frac{Sw - Swc}{1 - Swc - Sorw} \tag{2.40}$$

where the J-function is calculated using capillary pressure (Pc), interfacial tension (σ), contact angle (θ), permeability (k) and porosity (ϕ) as [Amyx, 1960]:

$$J = \frac{Pc}{\sigma Cos\,\theta}\sqrt{\frac{k}{\varphi}} \tag{2.41}$$

Figure 2.30 shows a typical plot of the normalized Leverett J-function as a function of normalized water saturation.

In addition to the Leverett J-function, a technique based on a modified Kozeny-Carman equation and the concept of mean hydraulic radius may be used for identification and characterization of flow units within a reservoir. The technique was proposed by Amaefule et al. [1993] and designated as the Flow Zone Indicator (FZI) technique, where the logarithm of $0.0314\sqrt{\frac{k}{\varphi}}$ versus the logarithm of $\phi/(1-\phi)$ would yield a straight line with unit slope for any flow unit.

2.3.2.6 Initialization data

Initialization of a model involves specification of pressures and saturations for each phase in each grid cell. Black-oil modeling of a reservoir can be initiated from the initial state when the reservoir was in capillary/gravity equilibrium. A proper initialization procedure should yield the correct fluids in place in each grid cell,

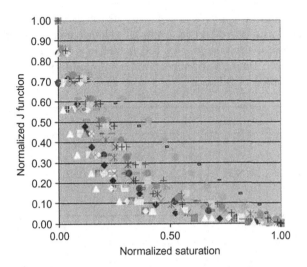

Figure 2.30 Typical plot of normalized Leverett J-function as a function of normalized water saturation.

maintain equilibrium conditions, and satisfy rock and fluid properties as well as the physical laws that govern the initial distribution of fluids in the reservoir [Aziz, Durlofsky, & Tchelepi, 2005].

If the capillary pressures that apply to the initial state of the reservoir are known, the pressures and saturations are determined by specifying one pressure (pressure at datum depth, P_{owc}) and two saturations (Sw at water-oil contact and Sg at gas-oil contact). The initial oil and water pressures are evaluated from the following equations:

$$P_o(z) = P_{owc} - \int_0^z \rho_o g \, dz \qquad (2.42a)$$

$$P_w(z) = P_{owc} - \int_0^z \rho_w g \, dz \qquad (2.42b)$$

Capillary pressures are defined when specifying the rock-fluid properties as functions of water saturations:

$$P_{cow}(Sw) = P_o - P_w \qquad (2.43)$$

The average initial saturation of water in a given grid cell i is [Aziz et al., 2005]:

$$Sw, i = \frac{ZSw_{i+1} - ZSw_i}{z_{i+1} - zi} \qquad (2.44)$$

where

$$ZSw = \int_0^z Sw \cdot dz \qquad (2.45)$$

The above technique is also applied for initialization of an oil-gas system.

For example, consider an oil reservoir with its top at a depth of 8325 ft, initial pressure of 4800 psia at reference depth of 8400 ft and water-oil contact of 8425 ft. Thickness of the reservoir is 100 ft and it is divided into three layers 20 ft, 30 ft and 50 ft thick. Variation of water-oil capillary pressure with water saturation is measured as follows:

Sw	Pc (psia)	Sw	Pc (psia)	Sw	Pc (psia)
0.12	40	0.32	4.001	0.62	0.872
0.121	35.919	0.37	2.793	0.72	0.5947
0.14	25.792	0.42	2.04	0.82	0.3317
0.17	18.631	0.52	1.555	1	0.1165
0.24	7.906	0.57	1.1655		

Density of fluids at standard conditions is reported as:

Oil density = 49.1 lb/ft^3
Water density = 64.79 lb/ft^3
Gas density = 0.06054 lb/ft^3

Base on the previously mentioned method, pressure and saturation for each grid cell are calculated as:

Layer No.	Thickness (ft)	Pressure (psia)	Water saturation
1	20	4783	0.187
2	30	4789	0.216
3	50	4800	0.311

Dividing the reservoir thickness (100 ft) into six layers of 10, 10, 15, 15, 25 and 25 ft and repeating the previous procedure yields the following initial pressure and saturation:

Layer No.	Thickness (ft)	Pressure (psia)	Water saturation
1	10	4781	0.181
2	10	4784	0.193
3	15	4787	0.207
4	15	4791	0.225
5	25	4796	0.265
6	25	4800	0.410

Figures 2.31 and 2.32 compare the results. It is obvious that refinement in the Z direction would generate more accurate data near the two-phase area (in this case, oil-water contact).

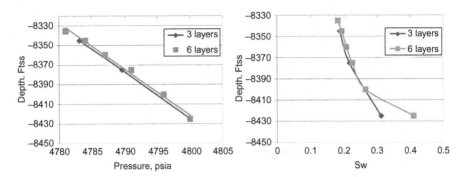

Figure 2.31 Initial pressure and water saturation for different layers.

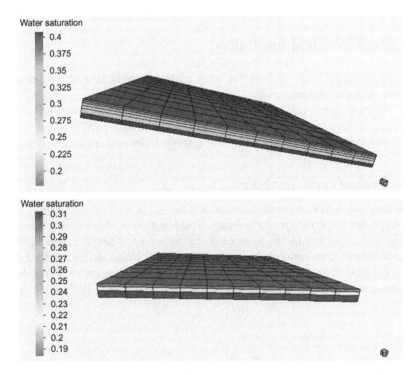

Figure 2.32 Initial water saturation for different layers (top for 5-layer discretization and bottom for 3-layer discretization).

2.3.2.7 Well and recurrent data

For dynamic reservoir modeling, well specifications such as well type (injector or producer), well coordinates, well trajectories, well completion intervals, and well constraints are required. In most cases, well pressures, production and injection histories are considered, too (Figure 2.33).

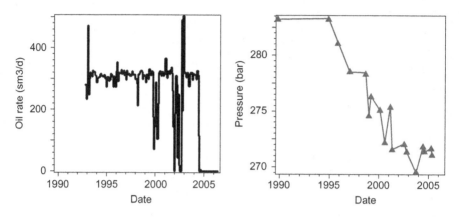

Figure 2.33 Typical oil rate and bottom-hole pressure data entered for reservoir modeling.

2.4 Mathematical modeling

Mathematical methods are perhaps the most common methods used by petroleum engineers. Material balance models, decline curve analysis models and well test models are three simple mathematical models that are widely used for reservoir characterization. Although nowadays some user-friendly software packages can be used in these simple models, hand calculations or graphical methods are also sufficient.

2.4.1 Decline curve analysis

Decline curve analysis is the most common method used in reservoir engineering calculations to predict future field production rates from historical production data. It involves matching production data history to an empirical equation and simply extrapolating the production trends. Conventional decline curve analysis, based on the work of Arps, matches the production rate versus time data to one of the following empirical decline curve equations: exponential, hyperbolic, or harmonic. The general form of the decline curve is:

$$
D_i = Kq^b = \frac{-dq/dt}{q} = -\frac{\Delta \ln q}{\Delta t}
$$
$$
K = \frac{D_i}{q_i^b}
$$

$$(2.46)$$

where for exponential, $b = 0$; for hyperbolic, $0 < b < 1$; and for harmonic, $b = 1$.

The decline rate, D_i, can be defined as the fractional change in the rate with time, Figure 2.34. If the production rate is plotted on a natural log scale against the time, the slope will be the decline rate, D_i, Figure 2.35.

After determination D_i, production data, usually flow rate versus time, is matched to one of the above empirical equations.

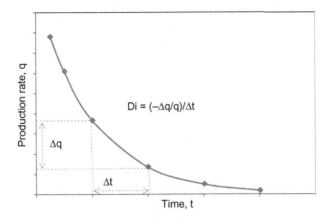

Figure 2.34 Rate of decline definition.

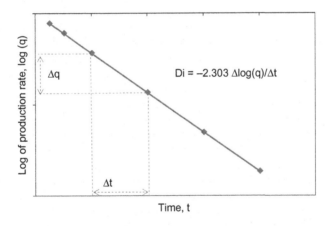

Figure 2.35 Rate of decline definition − a semilog plot.

The selected function is then applied to predict the rate in future, Figure 2.36. The implicit assumption in extrapolating the selected function is that the primary recovery mechanism remains unchanged in future. If the recovery mechanism changes, the method is not applicable.

2.4.2 Analytical models

Analytical models are based on the exact solutions of theoretical models, such as in pressure transient analysis and Buckley-Leverett models. The equations that govern physical behavior of the reservoirs are often complicated and, to solve them, some simplifications are applied. In fact, analytical models are the solution of simplified problems. For example, in pressure transient analysis we

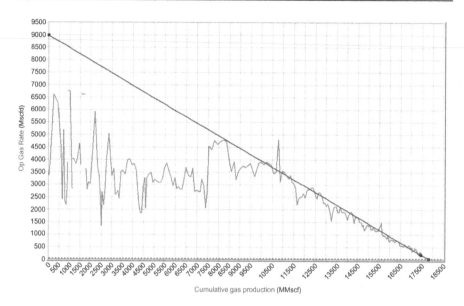

Figure 2.36 A field decline curve plot − Rate against cumulative production.

Figure 2.37 Horizontal block.

assume that the reservoir is horizontal, thickness is constant, fluid is single phase, flow regime is laminar and pressure gradient is small. The partial differential equation for one-dimensional fluid flow of a liquid into a porous medium with constant porosity (ϕ) constant permeability (k), viscosity (μ) and compressibility (c) is as follows to relate pressure (P) with time (t) and distance (x) (linear diffusivity equation):

$$\frac{\partial^2 P}{\partial x^2} = \left(\frac{\phi c \mu}{k}\right) \frac{\partial p}{\partial t} \tag{2.47}$$

For a horizontal block, Figure 2.37, with initial and boundary conditions of

$$P(x, 0) = P_i$$
$$P(0, t) = P_0$$
$$P(L, t) = P_L = P_i$$

the analytical solution of the transient pressure equation (Eq. 2.47) is given by [Kleppe, 2004]:

$$P(x,t) = P_0 + (P_L - P_0)\left[\frac{x}{L} + \frac{2}{\pi}\sum_{1}^{\infty}\frac{1}{n}\exp\left(-\frac{n^2\pi^2}{L^2}\frac{kt}{\phi\mu c}\right)\sin\left(\frac{n\pi x}{L}\right)\right]$$

The steady-state solution yields:

$$P(x,t) = P_0 + (P_L - P_0)\left(\frac{x}{L}\right) \tag{2.48}$$

Another example is the one-dimensional (along x axis) displacement process in which water with a flow rate of q_w and saturation of S_w displaces oil in a cubic porous medium with a cross-sectional area of A and porosity of ϕ (Buckley-Leverett problem). The continuity equation of water (with constant density) is written as:

$$-\frac{\partial q_w}{\partial x} = A\phi\frac{\partial S_w}{\partial t} \tag{2.49}$$

By definition of water fractional flow (f_w) as the ratio of water flow rate (q_w) to total flow rate (q):

$$f_w = \frac{q_w}{q}$$

the continuity equation is written as follows:

$$-\frac{\partial f_w}{\partial x} = \frac{A\phi}{q}\frac{\partial S_w}{\partial t} \tag{2.50a}$$

or

$$-\frac{df_w}{dS_w}\frac{\partial S_w}{\partial x} = \frac{A\phi}{q}\frac{\partial S_w}{\partial t} \tag{2.50b}$$

This equation is known as the Buckley-Leverett equation. A solution of the one-dimensional Buckley-Leverett equation by the analytical method of characteristics is shown in Figure 2.38.

Despite these simplified assumptions, the physics of the problem is always satisfied and one may study the effect of different parameters on the reservoir performance using the analytical models.

2.4.3 Numerical simulation

In numerical simulation the flow and heat equations of a constructed model are solved numerically. It is a blend of engineering, physics, chemistry, mathematics, numerical analysis, computer programming, and engineering experience and

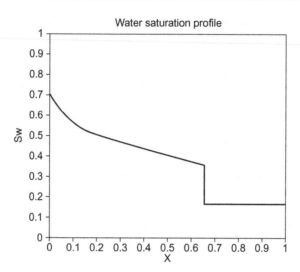

Figure 2.38 Solution of one-dimensional Buckley-Leverett equation.

practice. Today, numerical simulation is an essential and powerful tool for reservoir management [Aziz et al., 2005].

The equations of fluid flow through reservoirs are partial differential equations and computer programs solve the equations by numerical methods (usually by finite difference methods). For that purpose, reservoir volume is discretized to grid-blocks (or cells), flow equations (in general, nonlinear partial differential equations) are developed for discretized cells, and numerical methods are applied to solve the equations for each cell at every time step. In the finite difference technique, the nonlinear equations are first linearized and linear algebraic equations are then solved for saturation and pressure by Newtonian iteration methods. As the solutions of each cell at every time may depend on the solutions of neighboring cells, there are two schemes of solution: explicit and implicit.

In the explicit scheme, pressure and fluid saturation at a new time step (e.g., time step $n + 1$) is calculated using solutions of equations at the previous time step (n). Although the explicit scheme is very simple, due to a very restrictive stability requirement for the time step size it is not used practically.

In the implicit scheme, all unknown terms are calculated at the new time step ($n + 1$). This approach is unconditionally stable and the resulting linear algebraic equations build a system of equations that should be solved simultaneously. In order to reduce the number of equations, one may solve the pressure equation (results from flow equations) implicitly. After the pressure is obtained, we explicitly compute saturations. This technique is called IMPES (IMplicit Pressure Explicit Saturations). In order to be stable, IMPES method needs very small time step. This method is not suitable in cases where saturations changes are large during a time step [Dake, 1978].

Generally, in dynamic reservoir modeling, there are two kinds of modeling: black-oil and compositional modeling. A generalized black-oil model is characterized by three phases (oil phase, water phase and gas phase) and three components (oil, water and gas). In this modeling, mass transfer between water and oil is not considered (oil and water are immiscible). But the gas component can be dissolved in the oil and water phases. In addition, instantaneous thermodynamic equilibrium exists throughout the reservoir. This black-oil reservoir modeling is suitable for natural depletion as well as any processes (such as water injection or immiscible gas injection) where fluid compositions remain unchanged. Here, change in fluid properties (formation volume factors of fluids, gas oil ratio and viscosity) of each phase with pressure is considered.

In black-oil reservoir modeling, the densities of oil and gas phases at reservoir conditions can be expressed in terms of black-oil parameters as follows [Aziz et al., 2005]:

$$\rho_o = \frac{\rho_{o,sc}(1 + Rs)}{B_o} \tag{2.51}$$

$$\rho_g = \frac{\rho_{g,sc}(1 + Rv)}{B_g} \tag{2.52}$$

where Rv is volume of oil vaporized in gas and the subscript sc represents standard conditions.

By using Darcy's equation, black-oil fluid properties and continuity equation, the flow equations of oil, gas and water in x-direction become:

$$\frac{\partial}{\partial x}\left(\frac{k_o}{\mu_o B_o}\frac{\partial P_o}{\partial x}\right) - Q_o = \frac{\partial}{\partial t}\left(\frac{\phi S_o}{B_o}\right)$$

$$\frac{\partial}{\partial x}\left(\frac{kg}{\mu_g B_g}\frac{\partial P_g}{\partial x} + Rs\frac{k_o}{\mu_o B_o}\frac{\partial P_o}{\partial x}\right) - Q_g - RsQ_o = \frac{\partial}{\partial t}\left(\frac{\phi S_g}{B_g} + Rs\frac{\phi S_o}{B_o}\right) \tag{2.53}$$

$$\frac{\partial}{\partial x}\left(\frac{k_w}{\mu_w B_w}\frac{\partial P_w}{\partial x}\right) - Q_w = \frac{\partial}{\partial t}\left(\frac{\phi S_w}{B_w}\right)$$

where Q_o, Q_g and Q_w are oil, gas and water well rates.

To show an example of black-oil simulation, consider a reservoir with six vertically drilled wells in it. In this reservoir, the levels of gas-oil contact and water-oil contact are 2355 m and 2395 m, respectively. Pressure at gas-oil contact is reported as 23446 kPa. PVT properties are tabulated in Table 2.3. To study the behavior of the reservoir, the reservoir is discretized to 2660 ($19 \times 28 \times 5$) cells. Porosity, horizontal permeability and vertical permeability of the reservoir are then generated by geostatistics (Gaussian technique) and shown in Figures 2.39 to 2.41.

Table 2.3 **PVT properties of the example**

P	Rs	B_o	B_g	μ_o	μ_g
kPa	m³/m³	m³/m³	m³/m³	cP	cP
4000	11.46	1.064	0.02908	4.338	0.0088
6000	17.89	1.078	0.01886	3.878	0.0092
8000	24.32	1.092	0.01387	3.467	0.0096
10000	30.76	1.106	0.01093	3.1	0.01
12000	37.19	1.12	0.00899	2.771	0.0104
14000	43.62	1.134	0.00763	2.478	0.0109
15000	46.84	1.141	0.00709	2.343	0.0111
16000	50.05	1.148	0.00662	2.215	0.0114
17000	53.27	1.155	0.0062	2.095	0.0116
18000	56.49	1.162	0.00583	1.981	0.0119
19000	59.7	1.169	0.00551	1.873	0.0121
20000	62.92	1.176	0.00521	1.771	0.0124
21000	66.13	1.183	0.00495	1.674	0.0126
22000	69.35	1.19	0.00471	1.583	0.0129
23000	72.57	1.197	0.00449	1.497	0.0132
23446	74	1.2	0.0044	1.46	0.0133
24500	80	1.22	4.19E-03	1.4	1.35E-02

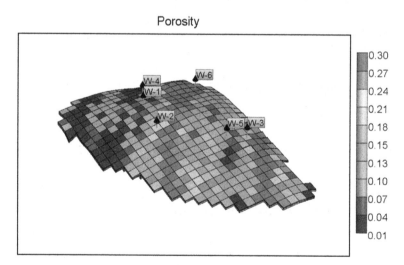

Figure 2.39 Generated porosity by geostatistics.

Wells 1, 2 and 5 were completed in layers 4 and 5, wells 3 and 4 in layers 3 and 4, and well 6 was completed in layer 4. Reservoir initial pressure and oil saturation are depicted in Figure 2.42. Relative permeability data for oil-water and gas-oil are shown in Figure 2.43. Capillary pressures are set to zero. It is desired to

Horizontal permeability

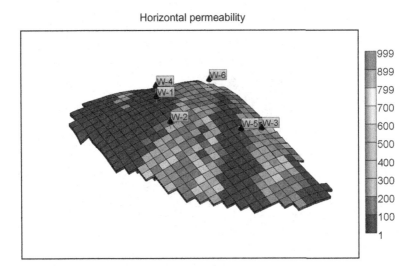

Figure 2.40 Generated horizontal permeability by geostatistics.

Vertical permeability

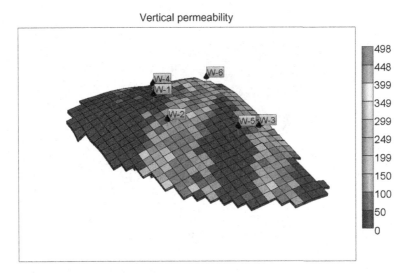

Figure 2.41 Generated vertical permeability by geostatistics.

predict pressure and oil saturation after 8.5 years production. The wells are produced under constraint of minimum bottom-hole pressure of 12000 kPa. The flow equations are then solved by the finite difference method and pressure and saturation of fluids are solved by the implicit scheme. The results are shown in Figures 2.44 and 2.45.

In the most enhanced oil recovery processes, such as miscible gas injection, fluid compositions change and black-oil modeling is not able to model them. Change in fluid

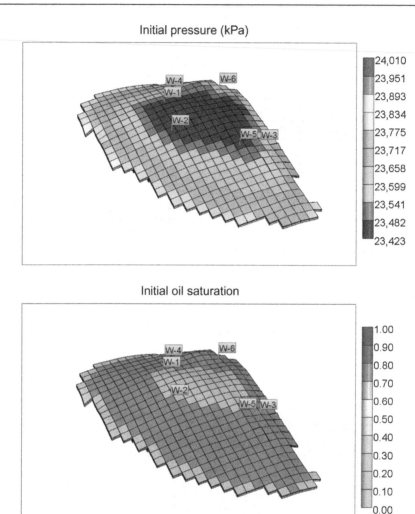

Figure 2.42 Initial pressure and oil saturation of the example.

composition should be considered in the modeling of depletion of gas condensate reservoirs as well as volatile oil reservoirs. In such cases, reservoir modeling is done by compositional models where oil and gas contain N components and water is a single-component material.

The equation of state is applied to perform flash calculation and to calculate fluid properties at a given temperature and pressure; Figure 2.46 shows typical PVT data for compositional modeling. In compositional models a flash calculation must be performed at each time-step in every grid cell. So, the time for running compositional models is much more than black-oil models.

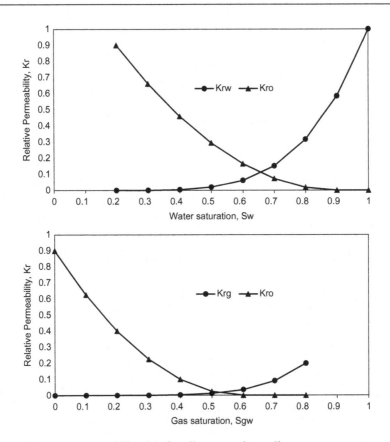

Figure 2.43 Relative permeability data for oil-water and gas-oil.

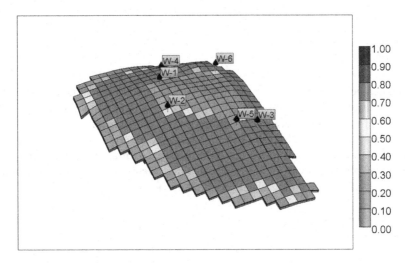

Figure 2.44 Oil saturation after 8.5 years production.

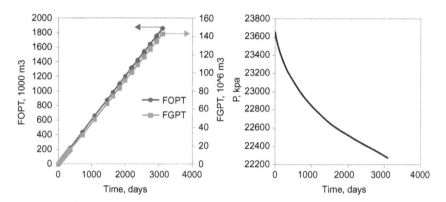

Figure 2.45 Results of numerical simulation.

The governing equation in compositional modeling is as follows [Aziz et al., 2005]:

$$\frac{\partial}{\partial x}\left(C_{ig}\rho_g\frac{k_g}{\mu_g}\frac{\partial P}{\partial x}+C_{io}\rho_o\frac{k_o}{\mu_o}\frac{\partial P}{\partial x}\right)+\frac{\partial}{\partial y}\left(C_{ig}\rho_g\frac{k_g}{\mu_g}\frac{\partial P}{\partial y}+C_{io}\rho_o\frac{k_o}{\mu_o}\frac{\partial P}{\partial y}\right)$$

$$+\frac{\partial}{\partial z}\left(C_{ig}\rho_g\frac{k_g}{\mu_g}\frac{\partial P}{\partial z}+C_{io}\rho_o\frac{k_o}{\mu_o}\frac{\partial P}{\partial z}\right)=\frac{\partial}{\partial t}\left[\varphi(C_{ig}\rho_gS_g+C_{io}\rho_oS_o)\right] \tag{2.54}$$

C_{ig} is mass fraction of component i present in the gas phase and C_{io} is mass fraction of component i present in the oil phase and

$$\sum_{i=1}^{N_c}C_{ig}=1 \tag{2.55}$$

$$\sum_{i=1}^{N_c}C_{io}=1 \tag{2.56}$$

The number of equations that must be solved in compositional modeling depends on the number of components. Often, we model the lighter components individually, and group heavier components into a pseudo-component. If nonhydrocarbons are involved, these may also have to be modeled separately.

In flash calculations, using an equation of state (such as Peng-Robinson), mole fractions of components in oil and gas phases are calculated. To do that, the following objective function is minimized for all components:

$$\sum_{i=1}^{Nc}\frac{z_i(K_i-1)}{1+V(K_i-1)}=0 \tag{2.57}$$

where z_i is the overall mole fraction of feed, K is the K-value ($=y/x$), V is mole of the gas phase and subscript i is for component i.

```
-- PVT Properties section
-- Peng-Robinson EOS
EOS
PR    /

-- Component Name
CNAMES
CO2 C1 C2 C3 'C4+' /

-- Binary Interaction Coefficients
BIC
0.2266599
0.2014434      -0.0008224
0.0717683       0.0135731      .0054848
0.2078494       0.0979894      .0054848      0.0      /

-- Critical Properties
PCRIT
73.76 46.00 48.8 31.6 16.3 /

TCRIT
304.2 190.0 358.4 488.3 741.6 /

-- Molecular Weight
MW
44.01 16.2828 41.1450 78.9157   224.0000 /

-- Acentric Factor
ACF
0.2250000   0.0086486   0.1366165   0.2727326   0.5856574  /

-- Critical Z factor
ZCRIT
0.27490002   0.28971521   0.27810650   0.26666820   0.22416106 /

ZCRITVIS
0.1140665    0.3219673    0.1884522    0.5301314    0.2372103  /

-- Omega A
OMEGAA
0.44162560 0.50724604 0.38625979 0.51680264 0.28916907 /

-- Omega B
OMEGAB
0.06815855 0.09485338 0.06251260 0.03738579 0.06581219 /

-- Parachor
PARACHOR
79.7        77.536      141.415      250.576      676.333     /

-- Standard Conditions
STCOND
15.0 1.0 /

-- Oil Water Gas gravity
GRAVITY
45.5 1.01 0.7773 /

-- Reservoir temperature: Deg C
RTEMP
71.0 /
```

Figure 2.46 Typical PVT data for compositional modeling.

2.5 Model verification

It is important that the reservoir engineers, who use models and simulators for reservoir study, be sure that the solutions obtained by numerical methods (the output of reservoir simulators) are as close as possible to the analytical solutions of given theoretical mathematical models. There are many techniques for models verification. One way is to define problems with known analytical solutions (usually one-dimensional problems) and then to check the model results with the solutions. Another technique includes checking the results with the results of some benchmark problems. The Society of Petroleum Engineers (SPE) published a series of comparative solution projects for various problems. The problems were solved by different reservoir simulators and the solutions were compared to each other.

Experimental design in reservoir engineering

3

3.1 Introduction

As stated in the previous chapter, reservoir characterization is the major part of reservoir modeling. Nevertheless, because of reservoir complexity and limited information, reservoir characterization is not usually conducted completely and precisely. In other words, uncertainties always exist in reservoir characterization and it is impossible to deterministically define a reservoir [Zee Ma & La Pointe, 2011].

Schulze-Riegert and Ghedan [2007] argued three sources for uncertainties in reservoir modeling: measuring errors, mathematical errors and incompleteness of data. Almost all experimental and field data suffer from errors caused by measurement tools and human errors. These types of errors may be minimized when using modern and accurate tools as well as increasing human knowledge and experience. When geologists and/or engineers estimate some reservoir properties by applying mathematical models, uncertainties should be taken into consideration, because none of the mathematical models are perfect (there are some simplifications in mathematical models). Besides these errors mentioned, all required data for reservoir modeling are not available and therefore we imperatively estimate unobtainable data.

3.2 Errors in mathematical modeling

The most common mathematical formulations used in reservoir modeling are based on material balance and Darcy's law.

The material balance equation is as follows:

Initial fluids in reservoir − produced fluids = remaining fluids

or in mathematical formulation:

$$F = N(E_o + mE_g + E_{f,w}) + (W_i + W_e)B_W + G_iB_g \tag{3.1}$$

where F is total (oil, gas, and water) produced fluids:

$$F = N_P[Bo + (R_P - R_S)B_g] + W_PB_w \tag{3.2}$$

Expansion of oil and solution gas, E_o, can be written as:

$$E_o = (B_o - B_{oi}) + (R_{Soi} - R_S)B_g \tag{3.3}$$

Experimental Design in Petroleum Reservoir Studies. DOI: http://dx.doi.org/10.1016/B978-0-12-803070-7.00003-X

For gas cap, rock and water expansions the following equations are used:

$$E_g = B_{oi}\left(\frac{B_g}{B_{gi}} - 1\right) \tag{3.4}$$

$$E_{f,w} = -(1 + m)B_{oi}\left(\frac{C_r + C_w S_{wi}}{1 - S_{wi}}\right)\Delta P \tag{3.5}$$

where C denotes compressibility and m is the ratio of initial gas volume to initial oil volume.

Assumptions applied in the material balance equation are as follows [Islam et al., 2007]:

- Rock and fluid properties are constant
- Darcy's law is applied for flow through porous media
- Complete segregation among phases exists
- Reservoir geometric properties are known
- PVT data from PVT laboratory are applicable throughout the reservoir

It should be noted that the material balance equation is sensitive to measured reservoir pressure. So, this equation shouldn't be applied in pressure maintenance projects [Islam et al., 2007].

In Darcy's law there are some assumptions, too [Islam et al., 2007]:

- Single, homogeneous and Newtonian fluid
- No reaction between fluids and rock
- Laminar flow regime
- Independency of permeability with pressure, temperature and fluid type
- No electrokinetic effects

The equations resulting from application of material balance and Darcy's law are nonlinear equations that can be solved numerically. Among the most common numerical methods is the finite difference method, whereby, using Taylor's series, some terms are truncated and cause truncation errors.

Discretization in the continuity equation: The continuity equation for one-phase, three-dimensional flow in Cartesian coordinates is:

$$\frac{\partial}{\partial x}(\rho u) + \frac{\partial}{\partial y}(\rho v) + \frac{\partial}{\partial z}(\rho w) + \frac{\partial}{\partial t}(\rho \varphi) = 0 \tag{3.6}$$

where u, v, and w are velocities in x, y, and z directions, respectively, and the corresponding Darcy equations are:

$$u = -\frac{k_x}{\mu}\left(\frac{\partial P}{\partial x} - \rho g \frac{dD}{dx}\right)$$

$$v = -\frac{k_y}{\mu}\left(\frac{\partial P}{\partial y} - \rho g \frac{dD}{dy}\right) \tag{3.7}$$

$$w = -\frac{k_z}{\mu}\left(\frac{\partial P}{\partial z} - \rho g \frac{dD}{dz}\right)$$

The above continuity equation results in a partial differential equation of:

$$k_x \frac{\partial^2 P}{\partial x^2} + k_y \frac{\partial^2 P}{\partial y^2} + k_z \frac{\partial^2 P}{\partial z^2} = \phi \mu c \frac{\partial P}{\partial t}$$

which can be numerically solved using standard finite difference approximations for the derivative terms of $\frac{\partial^2 P}{\partial x^2}, \frac{\partial^2 P}{\partial y^2}, \frac{\partial^2 P}{\partial z^2}, \frac{\partial P}{\partial t}$.

The x, y, and z-coordinate must be subdivided into a number of several grid cells, and the time coordinate must be divided into discrete time steps. Then, the pressure in each cell can be solved for numerically for each time step.

By applying Taylor series to pressure functions, approximations to the derivatives are obtained.

At constant time, t, the pressure functions may be expanded by Taylor's series as:

$$P(x + \Delta x, t) = P(x, t) + P'(x, t) \cdot \Delta x + P''(x, t) \frac{(\Delta x)^2}{2} + P'''(x, t) \frac{(\Delta x)^3}{6} + \cdots$$

$$P(y + \Delta y, t) = P(y, t) + P'(y, t) \cdot \Delta y + P''(y, t) \frac{(\Delta y)^2}{2} + P'''(y, t) \frac{(\Delta y)^3}{6} + \cdots$$

$$P(z + \Delta z, t) = P(z, t) + P'(z, t) \cdot \Delta z + P''(z, t) \frac{(\Delta z)^2}{2} + P'''(z, t) \frac{(\Delta z)^3}{6} + \cdots$$

One may write first derivatives of pressure as forward discretization:

$$P'(x, t) = \frac{P(x + \Delta x, t) - P(x, t)}{\Delta x} + \varepsilon(\Delta x)$$

$$P'(y, t) = \frac{P(y + \Delta y, t) - P(y, t)}{\Delta y} + \varepsilon(\Delta y)$$

$$P'(z, t) = \frac{P(z + \Delta z, t) - P(z, t)}{\Delta z} + \varepsilon(\Delta z)$$

At constant time, t, the pressure functions may be also expressed as:

$$P(x - \Delta x, t) = P(x, t) - P'(x, t) \cdot \Delta x + P''(x, t) \frac{(\Delta x)^2}{2} - P'''(x, t) \frac{(\Delta x)^3}{6} + \cdots$$

$$P(y - \Delta y, t) = P(y, t) - P'(y, t) \cdot \Delta y + P''(y, t) \frac{(\Delta y)^2}{2} - P'''(y, t) \frac{(\Delta y)^3}{6} + \cdots$$

$$P(z - \Delta z, t) = P(z, t) - P'(z, t) \cdot \Delta z + P''(z, t) \frac{(\Delta z)^2}{2} - P'''(z, t) \frac{(\Delta z)^3}{6} + \cdots$$

which yield backward discretization of first derivatives:

$$P'(x,t) = \frac{P(x,t) - P(x - \Delta x, t)}{\Delta x} + \varepsilon(\Delta x)$$

$$P'(y,t) = \frac{P(y,t) - P(y - \Delta y, t)}{\Delta y} + \varepsilon(\Delta y)$$

$$P'(z,t) = \frac{P(z,t) - P(z - \Delta z, t)}{\Delta z} + \varepsilon(\Delta z)$$

Adding Taylor's series expressions, and solving for the second derivatives, we get the following approximation:

$$\left(\frac{\partial^2 P}{\partial x^2}\right)_i^t = \frac{P_{i+1}^t - 2P_i^t + P_{i-1}^t}{(\Delta x)^2} + \varepsilon(\Delta x^2)$$

$$\left(\frac{\partial^2 P}{\partial y^2}\right)_j^t = \frac{P_{j+1}^t - 2P_j^t + P_{j-1}^t}{(\Delta y)^2} + \varepsilon(\Delta y^2) \tag{3.8}$$

$$\left(\frac{\partial^2 P}{\partial z^2}\right)_k^t = \frac{P_{k+1}^t - 2P_k^t + k_{i-1}^t}{(\Delta z)^2} + \varepsilon(\Delta z^2)$$

ε denotes truncation error.

Truncation errors may be cut by decreasing grid sizes (Δx, Δy, Δz). However, round-off errors and numerical dispersion may arise. Chen [2007] defined numerical dispersion as the spreading of a flood front in a displacement process such as water flooding.

Numerical dispersion is usually expressed by

$$a \frac{\partial^2 P}{\partial x^2}$$

where a is defined as the numerical dispersion coefficient. It is severe when the physical diffusion coefficient (D in Fick's law, $D\frac{\partial P}{\partial x}$) is small compared to the numerical dispersion coefficient [Chen, 2007]. This condition usually happens in petroleum reservoirs.

Figures 3.1 and 3.2 show comparisons of exact solutions and approximation solutions for a simple case of:

$$\frac{\partial u}{\partial x} + \frac{\partial u}{\partial t} = 0 \tag{3.9}$$

As we see, the difference between exact solution and approximate solution (using finite difference) could be severe.

Grid orientation effects are another disadvantage of using finite difference methods in solving PDEs [Chen, 2007; Mattax, 1989]. These effects are severe when

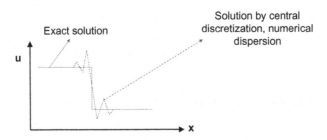

Figure 3.1 Error caused numerical dispersion.

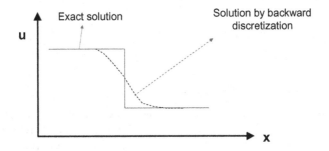

Figure 3.2 Solution of a PDE by finite difference method.

simulating cases with high mobility ratios (the ratio of k/μ of displacing phase to k/μ of displaced phase).

These all mean that, even if reservoir characterization is performed correctly and properly, forecasting results will be different from real results.

3.3 Uncertainty in reservoir data

In reservoir engineering, apart from outcrop and seismic data, all other data are gathered from points where wells intersect the reservoir. The volume of these points is less than a percent of reservoir volume. In addition, all reservoirs are heterogeneous and properties from a point to another point are different. Therefore, there are uncertainties in reservoir data (we are not 100% sure about the data), mainly those prepared for space between wells. It should be noted the degree of uncertainty is different for different types of data.

Generally, uncertainties of reservoir engineering data are classified into five groups [Schulze-Riegert and Ghedan, 2007]:

1. Uncertainty in geophysical data
2. Uncertainty in geological data
3. Uncertainty in dynamic data

4. Uncertainty in PVT
5. Uncertainty in field performance data (production and pressure)

These groups are discussed in the following sections [Schulze-Riegert and Ghedan, 2007].

3.3.1   Uncertainty in geophysical data

Seismic data, which are used for building reservoir structure and shape, suffer from uncertainties. These uncertainties correspond to data gathering, data processing and data interpretation. These uncertainties are because of the following:

* Data gathering error
* Different interpretations
* Depth data converting error
* Error in preliminary interpretation
* Error in wave length map corresponding to reservoir crest

3.3.2   Uncertainty in geological data

Perhaps the most uncertain data in reservoir modeling result from geological data. In geological data, uncertainties are about sedimentation, nature of rock (lithology), extension area of rock, and rock properties. These result in the following uncertainties:

* Uncertainty in gross volume of reservoir
* Uncertainty in direction and size of sedimentation
* Uncertainty in different rock type extension
* Uncertainty in porosity
* Uncertainty in net-to-gross ratio
* Uncertainty in fluid contacts

These uncertainties have some effects on evaluation of hydrocarbon in-place and dynamics of fluid flow through the reservoir.

3.3.3   Uncertainty in dynamic data

Uncertainties of all parameters that affect fluid flow through a reservoir (such as absolute permeability, vertical to horizontal permeability, relative permeability, fault transmissibility, injectivity, productivity index, skin, capillary pressure, aquifer properties) are considered in this section. These uncertainties influence both the reserve estimation and production profile.

3.3.4   Uncertainty in PVT data

PVT data could be considered the least uncertain data. Uncertainty in PVT data affects the capacity of processing units, hydrocarbon transport and marketing. Some of the uncertainties in this section are:

* Uncertainty in fluid samples: sampled fluid may be unrepresentative of the fluid reservoir. This would affect selected enhanced oil recovery (EOR) scenarios.
* Uncertainty in fluid composition

- Uncertainty in measured PVT properties
- Uncertainty in interfacial tension data

3.3.5 Uncertainty in field performance data

In addition to above uncertainties, field performance data may suffer uncertainties:

- Oil production rates are usually measured systematically and accurately but Water Oil Ratio (WOR) and Gas Oil Ratio (GOR) measurements are done only occasionally.
- Rate fluctuations are usually smoothed out as they can occur at short intervals.
- Gas rates are not measured accurately, especially when a part of it is flared.
- Injection data are less accurate than production data due to the measurement errors, fluid losses into other intervals caused by leaks in the casing or flow behind the pipe.
- Pressures measured during flow tests are usually less reliable than those obtained during shut in.

3.4 Uncertainty analysis

To reduce the effects of uncertainties in reservoir engineering data, and to quantify the uncertainties, an analysis of uncertainty is usually conducted. Among the advantages of uncertainty analysis in reservoir modeling, we can list the most important of them as follows:

- It helps reservoir managers in making their decisions.
- The effects of different factors (parameters) on the selected responses are studied (sensitivity studies).
- Results of different scenarios are easily compared.

3.4.1 History matching

As stated earlier, reservoir models (constructed at the final stage of the reservoir characterization phase) suffer from uncertainties in the data and in the modeling. Therefore, the true reservoir performance cannot be predicted by models. A standard method in a reservoir study is changing the reservoir properties so that field data and model results are matched. This method is called history matching. Once the model historically matches field data, one can say it will behave the same as the actual reservoir under future constraints within some accepted tolerance. The main purpose of history matching is tuning and the validity check of reservoir parameters. History matching is an iterative method, where model parameters are adjusted (tuned) using the following steps [Islam et al., 2007]:

1. Specifying the objective (which of the reservoir performances should be matched?)
2. Choosing the method of history matching (as there are different techniques for history matching)
3. Specifying the criteria for matching (accepted tolerance)
4. Specifying the parameters that should be tuned (usually the most influential parameters are selected)

5. Running the simulation and reporting the results
6. Comparing the results with field data (observed data)
7. Changing the reservoir parameters
8. Repeating steps 5 to 7

Variables that may be considered for a history match study could be:

- Porosity
- Water saturation
- Permeability
- Net thickness
- Vertical to horizontal permeability ratio
- Fault transmissibility
- Aquifer properties
- Pore volume
- Fluid properties
- Rock compressibility
- Relative permeability
- Capillary pressure
- Fluid contacts
- Well inflow parameters

A history matching phase usually starts with matching the total liquid production rate (this step is done simply as wells are constrained under total liquid rate). The next step is to compare average reservoir pressure by changing pore volume, permeability, compressibility and aquifer properties. After getting the reservoir pressure match, the next step is the saturation match. Rietz and Palke [2001] stated that the time required for history matching depends on years of historical data to match, number of wells, number of grid cells, number of phases, type of simulation (black-oil or compositional), scopes and objectives of the model.

History matching can be done manually or automatically. In manual history matching, the reservoir simulation engineer runs the model, compares model results with the field data and changes model parameters in order to have a history-matched model. This type of history matching is a trial and error process, at the end of which the difference between model and field data would be minimized.

For example, consider a reservoir model that contains 9000 grid cells and has estimated relative permeability data as shown in Figure 3.3. Based on the estimated relative permeability data, maximum oil relative permeability and water relative permeability are 0.5 while maximum gas relative permeability is 1. Cumulative oil, gas and water production of the field are to be matched (Figure 3.4). Results of the initially built model using the previously mentioned relative permeability data slightly differ from the field data (Figure 3.5). One way to improve the match is to change the endpoint relative permeability data as follows:

Maximum oil relative permeability = 0.58
Maximum water relative permeability = 0.60
Maximum gas relative permeability = 0.55

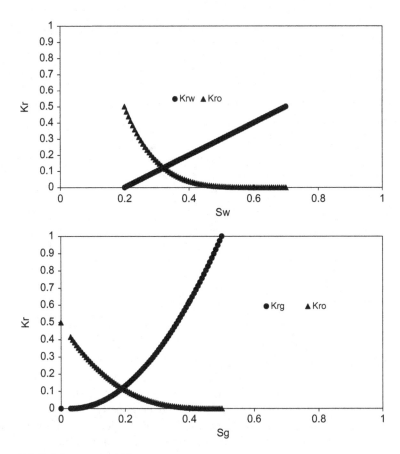

Figure 3.3 Relative permeability curves.

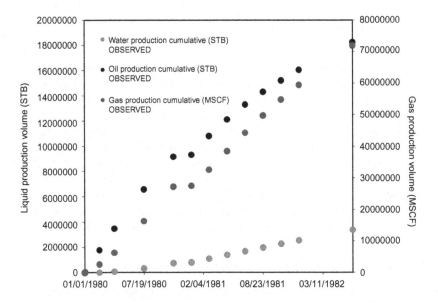

Figure 3.4 Field observed data.

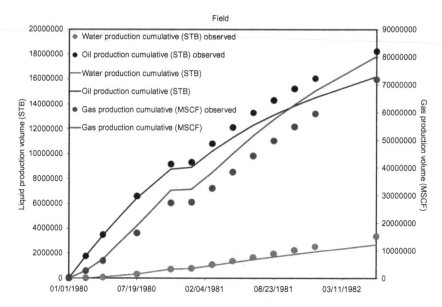

Figure 3.5 Model results compared with field observed data.

Figure 3.6 shows a comparison of the matched model with field data. It is clear that by these alterations, model matches the field data properly.

In an automatic history matching, by using computer logic, model parameters are adjusted to minimize the objective function (the function that compares model results and observed data):

$$F = \sum \left(Weight \left(\frac{observed - model}{StdDev} \right) \right)^2 \tag{3.10}$$

Automatic history-matching methods are classified into *deterministic* and *stochastic* methods. Deterministic methods are based on inverse problem theory while stochastic methods are a mimic of trial-and-error techniques. Among deterministic methods, the gradient-based method is the most common method, which attempts to minimize the gradient of the model objective with respect to the model parameters (porosity, permeability,...). Among stochastic methods, genetic algorithms and experimental design are the most efficient methods [Islam et al., 2007].

A history-matched model, however, does not guarantee a reasonable model, as more than one model with different parameters may match field observed data (because history matching is an inverse problem and so doesn't have a unique solution); these history-matched models result in different forecasts; see Figure 3.7.

3.4.2 Stochastic methods for uncertainty analysis

For a number of years, stochastic methods of Monte Carlo simulation and experimental design have been used for uncertainty analysis, in which probability

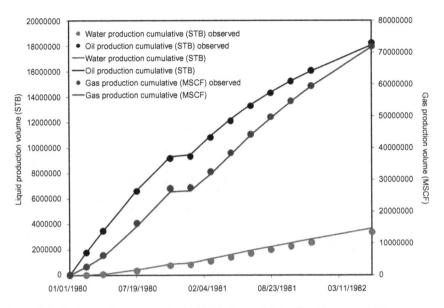

Figure 3.6 Model results compared with field observed data after history-matching.

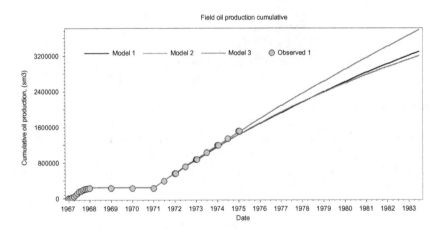

Figure 3.7 Different history-matched models results in different forecasts.

distributions are used to quantify uncertainties. In other words, probabilistic methods can be used as viable tools for uncertainty analysis and quantification [Deutsch and Journel, 1992].

In the next sections, these methods are explained in detail.

3.4.2.1 Basic definitions

Frequency Distribution: Frequency means how often something occurs. By counting all frequencies, frequency distributions are constructed. Although frequency

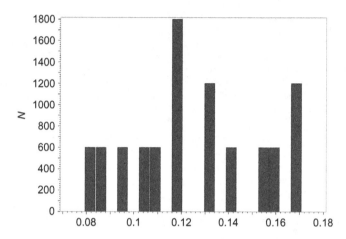

Figure 3.8 An example of histogram of porosity in a reservoir.

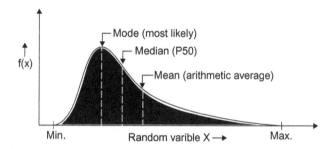

Figure 3.9 Mean, mode and median definition.

distributions are reported as tables or plots, graphical presentations are easier to understand. The most common graph is the histogram (Figure 3.8).

In statistics, data (or samples) are characterized by some useful information. This information is summarized as follows:

Mean: the arithmetic average of the data.
Median: the median is the sample point located at the center of the data.
Mode (most likely): the most frequently occurring point is called the mode. See Figure 3.9.
Variance: a measure of spreading in data. It is calculated as the average squared deviation of all data from the mean. Square root of the variance is called **standard deviation**.

Random Variables: A random variable is a variable that assigns a real value to the outcome of a random experiment. The value of a random variable is subject to variations due to chance, and is generated on the basis of some probabilistic functions. Although the outcome of any single random experiment related to the random variable cannot be predicted accurately, reliable prediction of the total results can be achieved by a great number of trials. The more trials there are, the more accurate the prediction will be [Sobol, 1974]. Random variables can be *discrete* or *continuous*.

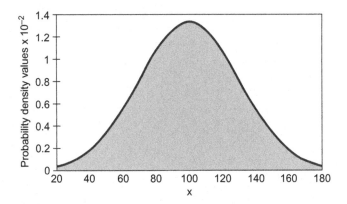

Figure 3.10 An example of probability distribution function (normal distribution).

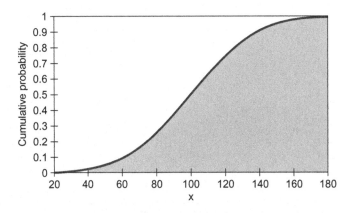

Figure 3.11 An example of cumulative distribution function.

Discrete random variable is a random variable that can take any of a specified finite or countable list of values x_i. The probability that a discrete random variable has the value x_i is equal to $P(x_i)$, and the sum of all probabilities is equal to one. Examples of discrete random variables in petroleum reservoir modeling are number of wells or number of rock types.

Continuous random variable is a random variable that takes any numerical value in an interval. In petroleum reservoir modeling, the possible values are infinite so there are many possible examples of continuous random variables such as porosity and fluid saturation.

Probability Distribution Function (PDF): A probability distribution function is a function that describes all the possible values that a random variable can take within a given range (Figure 3.10).

Cumulative Distribution Function (CDF): The probability that a random variable is less than a particular value. CDF is calculated by summing up the probabilities (Figure 3.11).

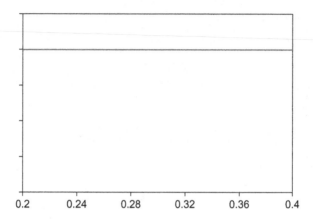

Figure 3.12 An example of uniform distribution.

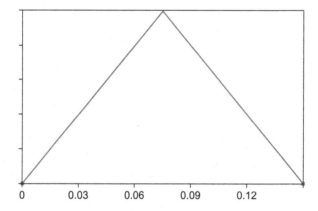

Figure 3.13 An example of triangular distribution.

Variable Distribution: In statistics, each random variable has a distribution. There are a few important distributions with specific characteristics. Among them, uniform, triangular, normal and log normal distributions are the most commonly used functions in statistics.

Uniform Distribution (a,b): The simplest distribution that returns a random number between "a" and "b". Figure 3.12 shows an example of uniform distribution for irreducible water saturation that varies between 0.2 and 0.4.

Triangular Distribution (a,b,c): A triangular-shaped distribution with minimum value of "a", median of "b" and maximum value of "c". Figure 3.13 is an example of a triangular distribution for critical gas saturation that has a minimum value of 0, median of 0.075 and maximum value of 0.15.

Normal Distribution (a,b): The normal distribution is the most common distribution in statistics. It returns a bell-shaped distribution with mean of "a" and

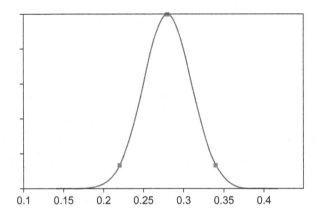

Figure 3.14 An example of normal distribution.

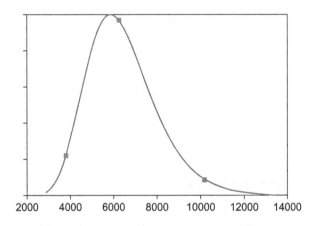

Figure 3.15 An example of log-normal distribution.

standard deviation of "b". Figure 3.14 shows a normal distribution of porosity with a mean value of 0.28 and standard deviation of 0.03.

Log-normal Distribution (a,b): In the log-normal distribution, the logarithm of the variable is normally distributed. Log-normal distribution is generated with a mean value of "a" and log of standard deviation equal to "b". A log-normal distribution of permeability with mean value of 8.74 and logarithm of a standard deviation of 0.25 is shown in Figure 3.15.

Scenario: A hypothesis to analyze possible events that can take place in the future. Each scenario involves the creation of a new object. Different production strategies are examples of scenarios in reservoir modeling.

Realization: Variation of parameters within a scenario is called realization. Different fluid contacts, petrophysical properties and fault transmissibility are examples of various realizations within a scenario (see Figure 3.16).

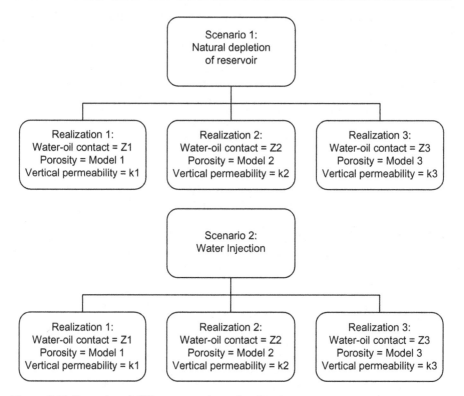

Figure 3.16 Examples of different scenarios and realizations.

3.4.3 Monte Carlo simulation

Monte Carlo is one of the most efficient stochastic methods and has been used for more than 50 years in different industries. In the petroleum industry, this method has been used for pressure transient analysis, reserve estimation, material balance analysis, risk evaluation, estimation of productivity and historical data [Baldwin, 1969; Gilman et al., 1998; Murtha, 1987, 1993; Wiggins and Zhang, 1993].

Monte Carlo simulations, in contrast to deterministic methods which give the same result with the same input data, give a distribution of result with useful information (optimistic case, most likely case and pessimistic case) about the dependent variable (Figure 3.17).

In fact, the Monte Carlo method is a mathematical model in which a dependent variable (such as hydrocarbon in-place, cumulative oil production) is a function of independent variables (such as porosity, permeability, saturation). Independent variables may have different distributions (normal, triangular, uniform,...). Distribution functions are determined by the use of different parameters. Random values of independent variables are then entered in the mathematical model and the dependent variable is calculated. This is repeated for thousands of iterations and, at the end, a distribution of the dependent variable is generated.

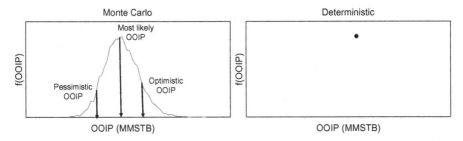

Figure 3.17 Monte Carlo simulation versus deterministic methods.

Table 3.1 Properties of reservoir for estimating OOIP

Property	Distribution function	Min. value	Mean Value	Max. value	Standard deviation
Porosity	Normal		0.14		0.02
Bulk Volume	LogNormal		900000 ft^3		2400 ft^3
Oil Formation Volume factor	Normal		1.34 rbbl/stb		0.06 rbbl/stb
Water Saturation	Triangle	0.2	0.3	0.45	

In reservoir engineering, one of the most common applications of Monte Carlo simulation is to evaluate original oil in place (OOIP) of a reservoir. OOIP is calculated using the following mathematical model:

$$OOIP = \frac{\varphi(1 - S_W)Ah}{Bo_i}$$

Here, dependent variable (OOIP) is related to independent variables of porosity (ϕ), water saturation (S_w), reservoir bulk volume (Ah) and initial oil formation volume factor (Bo_i). By using a program that generates random values (like Excel), random values are generated for all dependent variables. Then the OOIP is calculated by the mathematical model. This process is repeated thousands of times and, at the end, the cumulative distribution function and probability density distribution for OOIP are generated. It should be noted that cumulative and probability density distributions strongly depend on distribution functions of input factors (independent variables).

As an example, suppose we'd like to estimate OOIP of a reservoir with the properties shown in Table 3.1.

Figures 3.18 and 3.19 show distribution functions for the above properties. Generated distribution for OOIP estimated by Monte Carlo simulation after 10,000 simulation trials is shown in Figure 3.20. Based on the Monte Carlo simulation, an optimistic value for OOIP is 77324 MMTB, most likely value is 62600 MMSTB, and a pessimistic value is 49115 MMSTB.

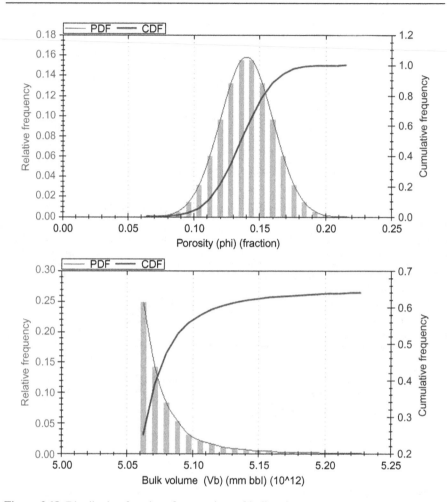

Figure 3.18 Distribution functions for porosity and bulk volume.

3.5 Experimental design

Reservoir engineers conduct different scenarios on reservoir models to discover the maximum possible economic recovery from a reservoir. So, testing different scenarios is a natural part of reservoir engineering and is an important part of the way we learn about how a reservoir behaves under different conditions. Each scenario usually involves various realizations and it is obvious that conducting all possible scenarios always requires time and cost. It is therefore desirable to obtain maximum information and knowledge by conducting a minimum of simulation runs at the lowest cost. Experimental design (or *design of experiments*) is a viable tool to increase our knowledge about the reservoir processes and to optimize the recovery processes at minimum cost and time. By using experimental design, realizations are designed

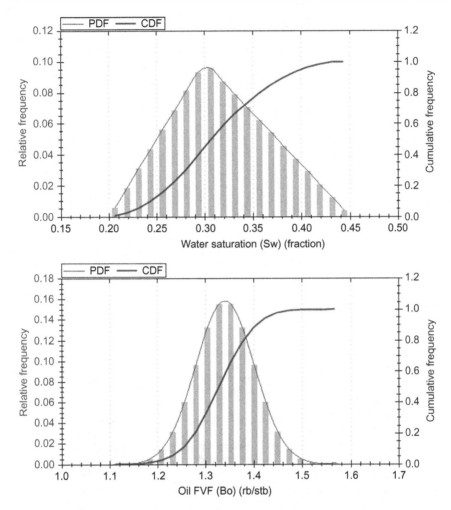

Figure 3.19 Distribution functions for water saturation and Oil FVF.

in a way that reliable, real and meaningful results are achieved, the results can be easily analyzed by stochastic methods and the reservoir study meets all specified objectives. If there are uncertainties in data or in a process, experimental design can be one of the most efficient techniques to quantify and to analyze the uncertainties.

Experimental design can also be widely used in reservoir studies where effects of one or more factors on a reservoir or a process are investigated. The general idea is based on the concept that each reservoir is controlled by some factors. Those factors influence the quality (or quantity) of produced products by the system (Figure 3.21). Experimental design enables reservoir engineers to control the reservoir and to adjust the most influential factors so that the hydrocarbon production is a maximum and the reservoir is at optimum conditions.

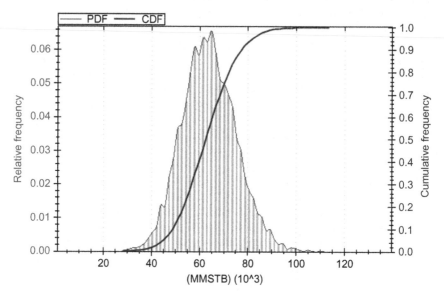

Figure 3.20 Monte Carlo results for OOIP estimation.

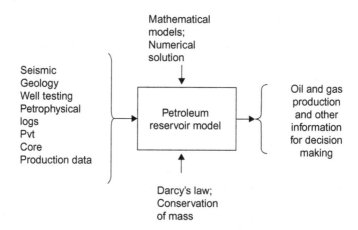

Figure 3.21 Input-output of a system.

One of the simplest methods to look at the effect of different parameters on a process or a reservoir is the approach of one-parameter-at-a-time (OPAT). In this method, realizations start with a starting point, or baseline set of levels, and then each parameter is successively varied over its range with the other parameters held constant [Montgomery, 2001]. After all realizations are conducted, the effect of varying each factor with all other factors held constant is analyzed. The major disadvantage of the OPAT strategy is that it cannot consider any possible interaction among the parameters [Montgomery, 2001].

Table 3.2 An example of one-parameter-at-a-time, OPAT

Factor name	Description	Low value	Mid. value	High value
AQPERM	Aquifer Permeability (md)	1	5.5	10
AQPOR	Aquifer Porosity	0.07	0.1	0.13
AQTHICK	Aquifer Thickness (ft)	400	500	600
SIGMA	Shape Factor $(ft)^{-2}$	0.02	0.04	0.06
FracPOR	Fracture Porosity	0.001	0.002	0.003
WOC	Water-oil contact (ft)	8400	8425	8450

Design	AQPERM	AQPOR	AQTHICK	SIGMA	FracPOR	WOC
Base Case	5.5	0.10	500	0.04	0.002	8425
Case 1	1	0.10	500	0.04	0.002	8425
Case 2	10	0.10	500	0.04	0.002	8425
Case 3	5.5	0.07	500	0.04	0.002	8425
Case 4	5.5	0.13	500	0.04	0.002	8425
Case 5	5.5	0.10	400	0.04	0.002	8425
Case 6	5.5	0.10	600	0.04	0.002	8425
Case 7	5.5	0.10	500	0.02	0.002	8425
Case 8	5.5	0.10	500	0.06	0.002	8425
Case 9	5.5	0.10	500	0.04	0.001	8425
Case 10	5.5	0.10	500	0.04	0.003	8425
Case 11	5.5	0.10	500	0.04	0.002	8400
Case 12	5.5	0.10	500	0.04	0.002	8450

Here is an example of OPAT: Consider studying a fractured petroleum reservoir with the six uncertain parameters of aquifer thickness (AQTHICK), aquifer porosity (AQPOR), aquifer permeability (AQPERM), shape factor (SIGMA), fracture porosity (FracPOR) and water-oil contact (WOC). Upper part of Table 3.2 shows the range for each factor. In OPAT design of N factors with 3 levels, the number of realizations is $3 + 2 * (N - 1)$. In this example, the number of factors is 6; so the number of realizations is 13. Lower part of Table 3.2 shows these realizations.

3.5.1 Basic rules in experimental design

In conducting a scenario, the reservoir engineer sets values of inputs and analyzes the response. Then he tries to explore how the response varies as he changes values of inputs. The scenario may also involve "nuisance" input variables (random error, measurement error, noise and bias) that affect the response. If the engineer is unable to completely control these additional variables, repeated scenarios at the same values of inputs result in different observations as these nuisance variables vary. In order to overcome this concern and to increase the validity of results, a variety of techniques are used. Replication, randomization and blocking are among these techniques. Replication means that the experimenter (reservoir engineer for example) conducts experiments (obtains observations) in a random sequence. By observing the response multiple times at the same set of inputs, the experimenter can estimate the magnitude and distribution of random error [Santner, Williams, & Notz, 2003]. By replication, the standard deviation of the results is reduced and, therefore, more accurate results are extracted [Antony, 2003]. Randomization is a process of conducting trials (realizations) randomly and helps to find changes in response as inputs vary. It is a method to reduce the effect of bias. By randomizing, the effects of noise factors on the results are eliminated. Blocking is a process to increase the accuracy of the design by lowering the effects of variations in noise factors (such as shift-to-shift, day-to-day or machine-to-machine) [Antony, 2003]. In blocking, experiments are conducted in relatively homogeneous sets called blocks [Santner, Williams, & Notz, 2003].

In reservoir modeling, repeated reservoir simulations yield identical results at the same set of inputs. Because of this, in applying experimental design in reservoir modeling studies, designs would not take more than one observation at any set of inputs, so none of the traditional principles of replication, randomization, and blocking are of use in reservoir modeling studies.

3.5.2   Outcomes of experimental design

Here are some outcomes when experimental design is used [Montgomery, 2001; Antony, 2003]:

- Recognition of input parameters and output results
- Determination of effects of input parameters on output results in a shorter time and lower cost
- Determination of the most influential factors
- Modeling and finding the relationship among input parameters and output results
- Improving the quality of products (results)
- Assisting in decision making in order to optimize the process and to improve the efficiency
- Exact determination of input parameters to have the minimum variation of output results
- Reduction in system variations by the exact control of the influential parameters
- Determination of the allowable ranges for input parameters
- Better understanding of the process and system performance

To use experimental design effectively, the following guidelines are recommended [Montgomery, 2001]:

Recognition of the uncertain parameters: As it is mentioned in Section 3.3, there are a variety of uncertain parameters in reservoir modeling. It helps to prepare a list of uncertain parameters that are to be studied by experimental design. To do this, a statement of the problem, main purpose of the problem and better understanding of the phenomena (by using analytical modeling) are often needed. Usually, one large comprehensive study (in which includes all the necessary facts, details and problems) is unable to answer all questions and it is more suitable to use a series of smaller studies in sequential studies. There is a golden rule in reservoir modeling that states defining a greater problem does not necessarily result in greater accuracy and reliability [Aziz et al., 2005].

Choice of factors ranges and levels: In studying uncertain parameters (factors), the reservoir engineer has to specify the range over which each factor varies. Wide range is recommended at the initial investigations (to screen the most influential factors). After selecting the most influential factors, the range of factor variations usually becomes narrower in the subsequent studies.

In studies conducted by experimental design, the primary factors are referred to as main effects (or main factors). Effects that reflect interactions between main effects are called interaction effects. Note that in some cases, where effects of main factors and their interaction effects cannot be considered independently, effects of main factors and their interactions are combined. In those cases the term *confounding* is applied and effects that are confounded are called *aliases* [Antony, 2003].

In addition to the range of factors, the levels that realizations will be run must be determined. If the objective of the study is just to identify the key factors in a minimum number of runs (screening), it is recommended to keep the number of factor levels low (two levels works well for screening) [Montgomery, 2001].

Selection of response variable: Response variables (dependent variables) are those that reflect the objective of the study. Selection of response variables should be done properly so that it provides useful information about the reservoir or the process. Usually, in a history-matching study *well water cut, well gas oil ratio* and *reservoir pressure* are response variables, while in a forecast study *cumulative oil and gas production* are also considered to be response variables.

Selection of design: There are several designs of experiments. In selecting the design, the objective of the study should be considered. The most appropriate designs are classical approaches [Antony, 2003]. Among classical approaches, full factorial design, fractional factorial design, Plackett-Burman design, and Box-Behnken design are reviewed and used in this book.

Statistical analysis of data: In experimental design, for analyzing the data and to obtain objective results and conclusions, statistical methods are employed. Analysis is usually done by a technique called analysis of variance (ANOVA) in which the differences between parameter means are analyzed. The technique is explained in the next sections.

Graphical analysis of data: Some simple graphical methods can be also used in the data analysis and interpretation. Some of these are main effects plots, interaction plots, cube plots, Pareto plots and tornado chart.

3.5.3 Designs

3.5.3.1 Two-level full factorial designs

One of the most common and powerful experimental designs is full factorial design, in which simulation runs are performed at all combinations of factor levels. Using full factorial designs, an engineer would be able to study the interaction effects of the factors on a response. The effect of each factor on the response variable is studied by changing the level of that factor. For example, if a reservoir engineer is interested in studying the effects of vertical permeability and porosity

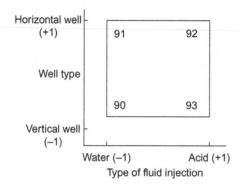

Figure 3.22 An example of two level factorial design.

on the cumulative oil production of a reservoir, and if two levels of vertical permeability (10 md and 22 md) and two levels of porosity (0.10 and 0.14) are considered important, a full factorial experiment would consist of making simulation runs at each of the four possible combinations of these levels of vertical permeability and porosity.

Among the most important factorial designs is that of k factors, each factor at two levels. The levels may be quantitative or qualitative. The complete design has 2^k runs and is called 2^k full factorial design. Full factorial design with two levels is a proper design for screening [Montgomery, 2001]. The 2^k full factorial design has k main effects, $\binom{k}{2}$ $(=k!/2(k-2)!)$ interaction effects between two factors and $\binom{k}{3}$ interaction effects between three factors and so on. The simplest full factorial design is 2^2 designs, including two factors each at two levels. The levels of each factor are low (or −1) and high (or +1).

As an example, consider a petroleum production engineer who seeks method(s) to raise oil production from wells located in the area of his supervision. He thinks the following factors are important:

- The type of well completion (open hole or cased hole)
- The type of well (horizontal or vertical)
- Type of well completion string (with and without tubing)
- The type of injection fluid (acid injection or hydraulic fracturing)

He then has four factors, each at two levels.

Suppose he initially considers two factors of "type of well" and "type of injection fluid." Figure 3.22 shows two level factorial designs for this case: a vertical well with hydraulic fracturing (or water injection) produces 90 bbl/day and with acid fracturing (or acid injection) produces 93 bbl/day, while a horizontal well that is hydraulically fractured produces 91 bbl/day and produces 92 bbl/day if it is stimulated by acid fracturing. The results may be shown as Table 3.3.

Table 3.3 Results of changing in fluid injection and well type

Type of fluid injection (design value)	Type of well (design value)	Oil production, bbl/day
Water (−1)	Vertical (−1)	90
Water (−1)	Horizontal (+1)	91
Acid (+1)	Vertical (−1)	93
Acid (+1)	Horizontal (+1)	92

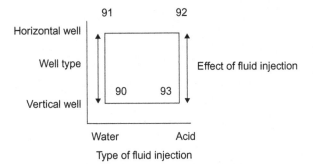

Figure 3.23 Effect of fluid injection.

The effect of "type of injection fluid" (Figure 3.23) is calculated as:

$$\frac{93 + 92}{2} - \frac{91 + 90}{2} = 2$$

It means switching from hydraulic fracturing to acid injection will raise production by 2 barrels of oil.

For calculating the effect of "type of well" (Figure 3.24) we can write:

$$\frac{91 + 92}{2} - \frac{90 + 93}{2} = 0$$

and a measure of interaction effect between "type of injection fluid" and "type of well" can be obtained by subtracting the average production rates on the right-to-left diagonal in the square from the average production rates on the left-to-right diagonal (Figure 3.25):

$$\frac{92 + 90}{2} - \frac{91 + 93}{2} = 1$$

The results of this process may be depicted as in Figure 3.26. It is obvious that the effect of "type of fluid injection" is larger than either "type of well" or the interaction.

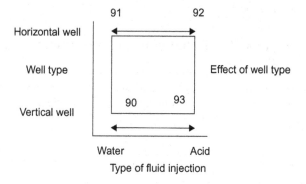

Figure 3.24 Effect of well type.

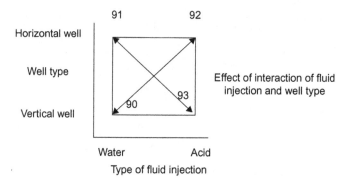

Figure 3.25 Effect of interaction of fluid and well type.

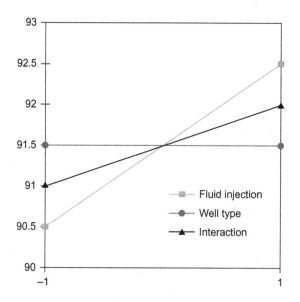

Figure 3.26 Main effects and interaction plot.

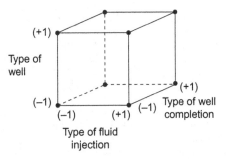

Figure 3.27 An example of 2^3 full factorial design.

Table 3.4 **An example of 2^3 full factorial design**

Type of fluid injection	Type of well	Type of fluid completion
Water (−1)	Vertical (−1)	Open hole (−1)
Water (−1)	Vertical (−1)	Cased hole (+1)
Water (−1)	Horizontal (+1)	Open hole (−1)
Water (−1)	Horizontal (+1)	Cased hole (+1)
Acid (+1)	Vertical (−1)	Open hole (−1)
Acid (+1)	Vertical (−1)	Cased hole (+1)
Acid (+1)	Horizontal (+1)	Open hole (−1)
Acid (+1)	Horizontal (+1)	Cased hole (+1)

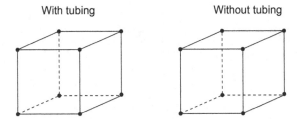

Figure 3.28 Design for all four factors.

Now, imagine the petroleum engineer would like to study the effect of three factors: "type of well," "type of fluid injection" and "type of well completion." Because each factor has two levels, a factorial design can be set up as shown in Figure 3.27 (cube plot). Notice that there are eight test combinations of these three factors across the two levels of each. These eight trials can be represented geometrically as the corners of a cube. This is an example of a 2^3 factorial design. This design is also shown in Table 3.4.

Figure 3.28 and Table 3.5 illustrate how all four factors—"type of well completion," "type of well," "type of fluid injection" and "type of completion string"—can be studied in a 2^4 factorial design. Because all four factors are at two levels, this experimental design can still be represented geometrically as a cube (actually a hypercube).

Table 3.5 **Full factorial design with 4 factors**

Type of fluid injection	Type of well	Type of fluid completion	Type of well completion string
Water (−1)	Vertical (−1)	Open hole (−1)	Without tubing (−1)
Water (−1)	Vertical (−1)	Open hole (−1)	With tubing (+1)
Water (−1)	Vertical (−1)	Cased hole (+1)	Without tubing (−1)
Water (−1)	Vertical (−1)	Cased hole (+1)	With tubing (+1)
Water (−1)	Horizontal (+1)	Open hole (−1)	Without tubing (−1)
Water (−1)	Horizontal (+1)	Open hole (−1)	With tubing (+1)
Water (−1)	Horizontal (+1)	Cased hole (+1)	Without tubing (−1)
Water (−1)	Horizontal (+1)	Cased hole (+1)	With tubing (+1)
Acid (+1)	Vertical (−1)	Open hole (−1)	Without tubing (−1)
Acid (+1)	Vertical (−1)	Open hole (−1)	With tubing (+1)
Acid (+1)	Vertical (−1)	Cased hole (+1)	Without tubing (−1)
Acid (+1)	Vertical (−1)	Cased hole (+1)	With tubing (+1)
Acid (+1)	Horizontal (+1)	Open hole (−1)	Without tubing (−1)
Acid (+1)	Horizontal (+1)	Open hole (−1)	With tubing (+1)
Acid (+1)	Horizontal (+1)	Cased hole (+1)	Without tubing (−1)
Acid (+1)	Horizontal (+1)	Cased hole (+1)	With tubing (+1)

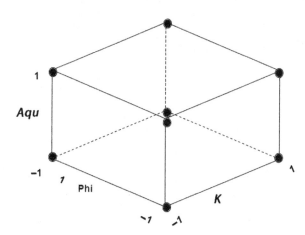

Figure 3.29 Low and high values of uncertain factors.

By convention, the low and high levels of a factor are denoted either by "−" and "+," respectively, or by "−1" and "+1". Figure 3.29 shows experimental design when effects of three factors, aquifer volume (Aqu), horizontal permeability (K) and porosity (Phi) are studied on a field performance.

3.5.3.2 Two-level fractional factorial designs

It is obvious as the number of factors in full factorial design rises, the number of realizations increases. For example, a complete 2^6 design requires 64 runs. In this

Table 3.6 Fractional factorial design, Resolution III

2^{3-1} design	A	B	C = AB
Run-1	−1	−1	+1
Run-2	+1	−1	−1
Run-3	−1	+1	−1
Run-4	+1	+1	+1

Table 3.7 Fractional factorial design, Resolution IV

2^3 design	A	B	C	D = ABC
Exp-1	−1	−1	−1	−1
Exp-2	+1	−1	−1	+1
Exp-3	−1	+1	−1	+1
Exp-4	+1	+1	−1	−1
Exp-5	−1	−1	+1	+1
Exp-6	+1	−1	+1	−1
Exp-7	−1	+1	+1	−1
Exp-8	+1	+1	+1	+1

design only 6 of the 64 runs correspond to the main effects, and 15 runs correspond to two-factor interactions. The remaining runs are associated with three-factor and higher interactions. The more runs that are done, the more budget and time required. If the engineers can reasonably assume that some higher-order interactions (e.g. third-order and higher) are not important, then information on the main effects and two-order interactions can be obtained by running only a fraction of the full factorial experiment. This design is called fractional factorial design. Fractional factorial designs are the most widely and commonly used types of design in industry. These designs are generally represented in the form $2^{(k-p)}$, where k is the number of factors and $1/2^p$ represents the fraction of the full factorial of 2^k. For example, $2^{(6-2)}$ is a {1/4} fraction of a 64 full factorial experiment. This means that one may be able to study 6 factors at two levels in just 16 runs rather than 64 runs.

The 2^{3-1} design is called a resolution III design, where a main effect is aliased with two-factor interactions (C = AB). Table 3.6 shows this type of design. In a resolution IV design, two-factor interactions are aliased with each other (D = ABC); see Table 3.7. Table 3.8 shows a resolution V for five factors where two-factor interactions are aliased with three-factor interactions [Montgomery, 2001].

3.5.3.3 Plackett-Burman design

Plackett-Burman design is one of the most commonly used of fractional factorial designs, as a standard two-level screening design [NIST Information Technology Laboratory, 2012]. It could be used for studying up to k = (N − 1)/(L − 1) factors, where L is the number of levels and N (a multiple of four) is the number of

Table 3.8 Fractional factorial design, Resolution V

2^4 design	A	B	C	D	E = ABCD
Exp-1	−1	−1	−1	−1	+1
Exp-2	+1	−1	−1	−1	−1
Exp-3	−1	+1	−1	−1	−1
Exp-4	+1	+1	−1	−1	+1
Exp-5	−1	−1	+1	−1	−1
Exp-6	+1	−1	+1	−1	+1
Exp-7	−1	+1	+1	−1	+1
Exp-8	+1	+1	+1	−1	−1
Exp-9	−1	−1	−1	+1	−1
Exp-10	+1	−1	−1	+1	+1
Exp-11	−1	+1	−1	+1	+1
Exp-12	+1	+1	−1	+1	−1
Exp-13	−1	−1	+1	+1	+1
Exp-14	+1	−1	+1	+1	−1
Exp-15	−1	+1	+1	+1	−1
Exp-16	+1	+1	+1	+1	+1

Table 3.9 Plackett-Burman design for 6 factors

DESIGN	A	B	C	D	E	F
DES-1	−1	−1	1	1	1	−1
DES-2	1	1	−1	1	−1	−1
DES-3	−1	1	−1	−1	−1	1
DES-4	1	1	−1	1	1	−1
DES-5	−1	−1	−1	−1	−1	−1
DES-6	−1	−1	−1	1	1	1
DES-7	1	−1	−1	−1	1	1
DES-8	1	−1	1	1	−1	1
DES-9	−1	1	1	−1	1	−1
DES-10	1	−1	1	−1	−1	−1
DES-11	−1	1	1	1	−1	1
DES-12	1	1	1	−1	1	1

simulation runs. In Plackett-Burman designs the importance of all main effects are at the same precision.

Table 3.9 presents rows of "+1" and "−1" values that are used to generate Plackett-Burman designs for N = 12 and k = 6.

3.5.3.4 Three-level designs

Usually, after identifying the key influential parameters of the study (screening), three-level designs are conducted at three levels. These levels are (usually) referred

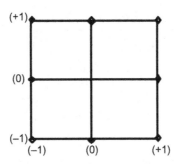

Figure 3.30 Three-level full factorial design for two factors.

to as low (−1), intermediate (0) and high (+1) levels. Where studying three-level designs, the effects of independent variables on the dependent variable are not linear [Montgomery, 2001].

Among three-level designs, full factorial design (3^k), fractional factorial design ($3^{(k-p)}$) and Box-Behnken are the most commonly used designs. Figure 3.30 shows three-level (−1, 0, +1) full factorial design for two factors. It is obvious that the number of runs becomes very large if there are several factors; for example, a five factor design at three levels involves 3^5 or 243 runs.

Box-Behnken is one of the fractional factorial three-level designs ($3^{(k-p)}$). It has been shown that these designs are efficient in the number of required runs [Montgomery, 2001]. Table 3.10 shows a three-level Box-Behnken design for six factors.

3.5.3.5 Latin hypercube design

A Latin hypercube (LH) of "n" runs for "m" parameters is an n × m matrix where each column consists of "n" equally spaced intervals [Lazić, 2004]. Each array of a Latin hypercube design, d_{ij}, in the design space of [0 1] is generated via

$$d_{ij} = \frac{l_{ij} + 0.5(n - 1) + u_{ij}}{n} ; i = 1, 2, ..., n; j = 1, 2, ..., m \tag{3.11}$$

where u_{ij} is an independent random number from [0 1] and l_{ij} is an array of the LH matrix. Table 3.11 shows an illustration of LH design for a reservoir study with five uncertain factors of critical water saturation (Swc), critical gas saturation (Sgc), vertical permeability multiplier (muk), aquifer size (Aqu), and aquifer productivity index (AqPI). Range of variation of each factor is also shown in the table. In this design, each uncertain factor is divided into 16 intervals. For convenience, minimum and maximum limits of factors are converted to "−1" and "+1".

3.5.4 Response surface

When the output of a system or a process is influenced by several variables (factors) and we want to optimize the output, the response surface method is a useful

Table 3.10 **Three-level Box-Behnken design for 6 factors**

Run no.	A	B	C	D	E	F
EX_001	1	0	1	0	0	−1
EX_002	0	−1	0	0	1	−1
EX_003	0	0	−1	1	0	−1
EX_004	0	0	1	−1	0	−1
EX_005	0	0	−1	−1	0	1
EX_006	1	0	0	−1	−1	0
EX_007	0	0	0	0	0	0
EX_008	1	0	0	1	1	0
EX_009	−1	−1	0	−1	0	0
EX_010	0	0	−1	−1	0	−1
EX_011	0	0	−1	1	0	1
EX_012	−1	0	0	−1	1	0
EX_013	−1	1	0	−1	0	0
EX_014	1	1	0	1	0	0
EX_015	1	1	0	−1	0	0
EX_016	0	0	1	1	0	−1
EX_017	1	0	0	1	−1	0
EX_018	0	−1	0	0	1	1
EX_019	0	1	−1	0	−1	0
EX_020	1	0	0	−1	1	0
EX_021	−1	0	−1	0	0	−1
EX_022	−1	0	1	0	0	−1
EX_023	0	0	0	0	0	0
EX_024	−1	0	1	0	0	1
EX_025	−1	1	0	1	0	0
EX_026	0	1	1	0	−1	0
EX_027	1	−1	0	−1	0	0
EX_028	0	1	−1	0	1	0
EX_029	−1	0	0	1	1	0
EX_030	0	0	0	0	0	0
EX_031	1	0	−1	0	0	1
EX_032	−1	0	−1	0	0	1
EX_033	0	0	1	−1	0	1
EX_034	1	0	1	0	0	1
EX_035	0	−1	−1	0	1	0
EX_036	0	1	0	0	1	1
EX_037	0	1	0	0	1	−1
EX_038	0	−1	−1	0	−1	0
EX_039	0	−1	0	0	−1	1
EX_040	0	1	0	0	−1	1
EX_041	1	0	−1	0	0	−1
EX_042	−1	0	0	−1	−1	0
EX_043	0	−1	1	0	1	0

(*Continued*)

Table 3.10 (Continued)

Run no.	A	B	C	D	E	F
EX_044	0	−1	1	0	−1	0
EX_045	0	−1	0	0	−1	−1
EX_046	0	1	1	0	1	0
EX_047	0	1	0	0	−1	−1
EX_048	1	−1	0	1	0	0
EX_049	−1	−1	0	1	0	0
EX_050	−1	0	0	1	−1	0
EX_051	0	0	1	1	0	1

Table 3.11 An example of Latin hypercube design

Factor	Min. value	Mid. value	Max. value
Critical Water Saturation (Swc)	0.22 (−1)*	0.35 (0)	0.48 (+1)
Critical Gas Saturation (Sgc)	0.04 (−1)	0.08 (0)	0.12 (+1)
Permeability Multiplier (muk)	0.001 (−1)	0.01 (0)	0.1 (+1)
Aquifer Size (Aqu)	9E8 (−1)	9E9 (0)	9E10 (+1)
Aquifer Productivity Index (AqPI)	1500 (−1)	5000 (0)	15000 (+1)

Design				
Swc	**Sgc**	**muk**	**Aqu**	**AqPI**
0.30 (−0.375)	0.12 (1)	0.042 (0.625)	5.06E9 (−0.25)	2667 (−0.5)
0.24 (−0.875)	0.06 (−0.5)	0.056 (0.75)	1.2E10 (0.125)	1500 (−1)
0.25 (−0.75)	0.08 (−0.125)	0.001 (−0.875)	2.85E9 (−0.5)	6325 (0.25)
0.27 (−0.625)	0.09 (0.25)	0.004 (−0.375)	9E10 (1)	5478 (0.125)
0.42 (0.5)	0.12 (0.875)	0.007 (−0.125)	1.6E9 (−0.75)	3080 (−0.375)
0.48 (1)	0.07 (−0.375)	0.006 (−0.25)	3.8E10 (0.625)	1732 (−0.875)
0.38 (0.25)	0.06 (−0.625)	0.100 (1)	3.8E9 (−0.375)	11248 (0.75)
0.37 (0.125)	0.11 (0.75)	0.032 (0.5)	6.75E10 (0.875)	9741 (0.625)
0.35 (0)	0.08 (0)	0.010 (0)	9E9 (0)	4743 (0)
0.40 (0.375)	0.04 (−1)	0.002 (−0.625)	1.6E10 (0.25)	8435 (0.5)
0.46 (0.875)	0.10 (0.5)	0.002 (−0.75)	6.75E9 (−0.125)	15000 (1)
0.45 (0.75)	0.09 (0.125)	0.075 (0.875)	2.85E10 (0.5)	3557 (−0.25)
0.43 (0.625)	0.07 (−0.25)	0.024 (0.375)	9E8 (−1)	4108 (−0.125)
0.29 (−0.5)	0.05 (−0.875)	0.013 (0.125)	5.06E10 (0.75)	7305 (0.375)
0.22 (−1)	0.10 (0.375)	0.018 (0.25)	2.13E9 (−0.625)	12989 (0.875)
0.32 (−0.25)	0.11 (0.625)	0.001 (−1)	2.13E10 (0.375)	2000 (−0.75)
0.33 (−0.125)	0.05 (−0.75)	0.003 (−0.5)	1.2E9 (−0.875)	2310 (−0.625)

*Values in parenthesis are design values.

tool for finding the relationship between response and variables. The method is a combination of mathematics and statistics that optimizes the output by a simple empirical model.

In most cases, we don't know the form of relationship between the response and the independent variables. So the first step is to find a suitable approximation for the true functional relationship between the response surface and the set of independent variables.

To find a suitable relationship, usually, a linear or second order polynomial of the independent variables is used as a starting point. If the response is well modeled by a linear function of the independent variables, then the approximating function is:

$$y = \beta_0 + \beta_1 X_1 + \beta_2 X_2 + \beta_3 X_3 + ...\beta_k X_k + \varepsilon \qquad (3.12)$$

Second-order response surfaces with consideration of interactions are:

$$y = \beta_0 + \sum \beta_i x_i + \sum \beta_{ii} x_{ii}^2 + \sum \sum \beta_{ij} x_{ij} \qquad (3.13)$$

Coefficients of above response surfaces (β_k) are found by applying regression techniques such as the least-squares technique. By applying the least-squares technique, we can fit a response surface containing independent variables by minimizing the residual error measured by the sum of squared deviations between the actual and the estimated responses.

It should be noted that a model with several coefficients (β_k) is not necessarily the best, and it is possible to improve a model significantly with a few coefficients by adding another coefficient [Steppan, Werner, & Yeater, 1998].

The test for significance of the regression model is done by comparing the effect caused by the regression model to the overall error. This comparison is based on the total sum of squares (SYY), the regression sum of squares (SSR), and the sum of squared errors (SSE):

$$SYY = \sum_{i=1} (Y_i - \overline{Y})^2 \qquad (3.14)$$

$$SSR = \sum_{i=1} (y_i - \overline{y})^2 \qquad (3.15)$$

$$SSE = \sum_{i=1} (y_i - Y_i)^2 \qquad (3.16)$$

where Y_i is a simulation output, $\overline{Y}$ is the average of all simulation outputs, y_i is an estimated output by response surface and $\overline{y}$ is the average of all estimated outputs.

The ratio between the Regression Sum of Squares (SSR) and the Sum of Squared Errors (SSE) is calculated and then is compared to the F ratio:

$$F = \frac{\left(\dfrac{SSR}{k}\right)}{\left(\dfrac{SSE}{n-p}\right)} = \frac{MSR}{MSE} \qquad (3.17)$$

where n is the number of points and p is the number of coefficients in the response model. MSR and MSE are regression mean square error and mean square error.

The calculated F ratio is then compared with standard F ratios (see Appendix). It is often recommended to provide a significance level, α; $100(1 - \alpha)$ is called a percent confidence interval, meaning that we are $100(1 - \alpha)$ percent sure that the groups are not equivalent [Gad, 2006]. In Appendix A, $\alpha = 0.05$. If the calculated F ratio for a factor (parameter) is greater than the value in Appendix A, the effect of that factor should be considered as an important factor. This method is called the F-test [Montgomery, 2003].

After determination of the response surface model, we must figure out if the model describes our data adequately. In order to check model accuracy, a common method is to use the coefficient of determination (R^2): the ratio of the regression sum of squares (SSR) over the total sum of squares (SYY). This coefficient ranges between 0 and 1. However, an R^2 value close to unity does not necessarily guarantee a good model because it is always possible to increase R^2 by adding higher order terms to the model equation, regardless of the significance of the terms added to the model [Steppan et al., 1998]. To overcome this concern, the adjusted coefficient of determination ($R^2_{adjusted}$) is applied:

$$R^2_{adjusted} = 1 - \frac{MSE}{\left(\dfrac{SYY}{n-1}\right)} \tag{3.18}$$

where n denotes the number of data points. When checking models the "best" model would be the one with the highest $R^2_{adjusted}$ (or with the lowest MSE).

When regressing, no linear relationship should exist between the independent variables. However, sometimes there are hidden relationships between variables. Relationships between variables can cause a problem called multicolinearity. In multicolinearity, estimation of coefficients can become unstable due to an increase in the variances of the coefficients, and the model can become inaccurate [Steppan et al., 1998]. To check for multicolinearity, one can use variance inflation factors (VIF):

$$VIF = \frac{1}{R^2} \tag{3.19}$$

If VIF for one of the variables is near or greater than 5, there is colinearity associated with that variable. If there are two or more variables that have a VIF around or greater than 5, one of these variables must be removed from the regression model [Steppan et al., 1998].

This method is explained by an example, as follows.

Suppose a reservoir engineer would like to find a response surface describing total field gas production (FGPT) as a function of six uncertain factors of water relative permeability at maximum water saturation (Krw), oil relative permeability at connate water saturation (Kro), gas relative permeability at connate water saturation (Krg), oil-water contact (WOC), horizontal permeability multiplier (Kx) and vertical permeability multiplier (Kz). He selects a Plackett-Burman design with a center

Table 3.12 **Range of variations of uncertain factors**

Parameter	Symbol	Low value	Mid. value	High value
Water relative permeability at maximum water saturation	Krw	0.2	0.6	1
Oil relative permeability at connate water saturation	Kro	0.2	0.6	1
Gas relative permeability at connate water saturation	Krg	0.2	0.6	1
Water-Oil contact	WOC	9900	9950	10000
Horizontal permeability multiplier	Kx	0.1	1	10
Vertical permeability multiplier	Kz	0.1	1	10

Run no.	Krw	Kro	Krg	WOC	Kz	Kx
1	0.20	0.20	1.00	10000	10	0.1
2	1.00	1.00	0.20	10000	0.1	0.1
3	0.20	1.00	0.20	9900	0.1	10.0
4	1.00	1.00	0.20	10000	10	0.1
5	0.20	0.20	0.20	9900	0.1	0.1
6	0.60	0.60	0.60	9950	1	1.0
7	0.20	0.20	0.20	10000	10	10.0
8	1.00	0.20	0.20	9900	10	10.0
9	1.00	0.20	1.00	10000	0.1	10.0
10	0.20	1.00	1.00	9900	10	0.1
11	1.00	0.20	1.00	9900	0.1	0.1
12	0.20	1.00	1.00	10000	0.1	10.0
13	1.00	1.00	1.00	9900	10	10.0

point for his study. Table 3.12 shows the design. The simulation results of FGPT for 13 simulation runs are shown in Table 3.13.

First, he chooses a linear relationship between the response surface (FGPT) and the six factors and tries to find the relationship as:

$$\text{FGPT} = \beta_0 + \beta_1 * \text{Krw} + \beta_2 * \text{Kro} + \beta_3 * \text{Krg} \\ + \beta_4 * \text{WOC} + \beta_5 * \text{Kz} + \beta_6 * \text{Kx} \tag{3.20}$$

He applies least-squares technique to determine regression coefficients (β_i). The results are shown in Table 3.14 and Figure 3.31. The significance level (α) is 0.05; so any variables with significance value higher than 0.05 are not significant. Looking at Table 3.14 reveals that regression coefficients of $\beta1$, $\beta2$, $\beta3$ and $\beta4$ have higher significance value. So, he can drop them and do regression without these variables (Krw, Kro, Krg, WOC):

$$\text{FGPT} = \beta_0 + \beta_1 * \text{Kz} + \beta_2 * \text{Kx} \tag{3.21}$$

Table 3.13 **Total field gas production (FGPT)**

Run no.	FGPT (MMSCF)
1	28.17205
2	24.9201
3	64.6562
4	36.71844
5	11.71267
6	118.0486
7	175.8971
8	171.8247
9	183.5038
10	82.88578
11	16.07094
12	112.399
13	173.6727

Table 3.14 **Results of linear regression**

Coefficient	Value	VIF	Significance
$\beta 0$	92.34		6.07065E-05
$\beta 1$	10.92	1.000	0.303038057
$\beta 2$	-7.661	1.000	0.459335046
$\beta 3$	9.248	1.000	0.376795639
$\beta 4$	3.399	1.000	0.737777178
$\beta 5$	21.33	1.000	0.070050174
$\beta 6$	56.79	1.000	0.001091388

ANOVA					
Source	*SS*	*SS%*	*MS*	*F*	*F Signif*
Regression	47456.9	88	7909.5	7.018	0.01595
Residual	6762.1	12	1127.0		
Total	54219.0	100			

Summary	
IR	0.936
R^2	0.875
R^2 adjusted	0.751
Collinearity	1.000

Then he tries to find the coefficients of the new relationship (Eq. 3.21). Least-squares regression is performed again. Table 3.15 and Figure 3.32 show the results of this regression. Comparing $R^2_{adjusted}$ in these two models (Eqs. 3.20 and 3.21) depicts that the second model is slightly better than the first model.

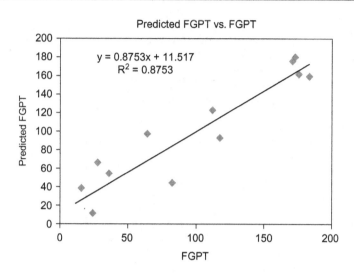

Figure 3.31 Predicted FGPT versus FGPT when using linear regression.

Table 3.15 **Results of linear regression, second model**

Coefficient	Value	Significance	VIF
β0	92.34	1.01735E-06	
β1	21.33	0.04212	1.000
β2	56.79	0.000101	1.000
ANOVA			

Source	SS	SS%	MS	F	F Signif
Regression	44157.9	81	22079.0	21.94	0.000220
Residual	10061.1	19	1006.1		
Total	54219.0	100			

Summary					
IR					0.902
R²					0.814
R² adjusted					0.777
Collinearity					1.000

In order to increase model quality, the reservoir engineer decides to use linear regression with some interactions of variables as follows:

$$\text{FGPT} = \beta_0 + \beta_1 * \text{Kx} + \beta_2 * \text{Kro} * \text{Kx} + \beta_3 * \text{Krw} * \text{Kx} + \beta_4 * \text{Krg} * \text{WOC}$$

$$(3.22)$$

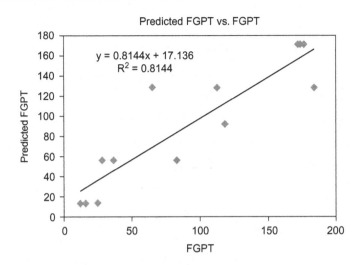

Figure 3.32 Predicted FGPT versus FGPT when using linear regression, second model.

Table 3.16 Results of linear regression with interaction of variables

Coefficient	Value	Significance	VIF
$\beta 0$	92.34	8.00314E-08	
$\beta 1$	63.00	3.78058E-06	1.167
$\beta 2$	−16.21	0.02097	1.167
$\beta 3$	24.64	0.002429	1.167
$\beta 4$	−18.63	0.01975	1.500

ANOVA					
Source	*SS*	*SS%*	*MS*	*F*	*F Signif*
Regression	51585.4	95	12896.4	39.18	2.67511E-05
Residual	2633.6	5	329.20		
Total	54219.0	100			

Summary	
IR	0.975
R^2	0.951
R^2 adjusted	0.927
Collinearity	0.667

Results of the least-squares method are shown in Table 3.16 and Figure 3.33. It is obvious that in this case, considering interaction of the parameters improves relationship between FGPT and uncertain factors.

Figure 3.34 illustrates a 3D response surface of FGPT as a function of Kx and Krw.

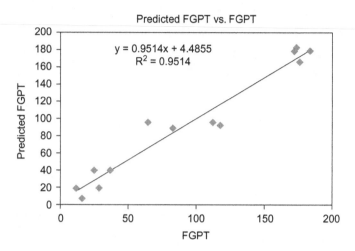

Figure 3.33 Predicted FGPT versus FGPT when using linear regression with interaction of variables.

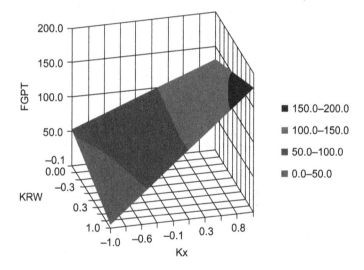

Figure 3.34 Three-dimensional plot of response surface.

3.5.5 Sensitivity analysis

In reservoir modeling we consider several input factors that we think they affect the reservoir performance (response). However, after analyzing the response, we may see that some of the inputs have little effect on the response. It means that the response is not sensitive to changes in these inputs. Hence, one can say that in sensitivity analysis of a reservoir study, input factors with most and little effect on the reservoir performance are determined. It is obvious that sensitivity analysis provides a better understanding of how different parameters affect the results. The information gained from sensitivity analysis is then used in other tasks such as in history matching and in predicting future performance of the reservoir. It also helps determine which factors should be varied and their approximate ranges. By sensitivity analysis one can identify whether there are interactions between input factors. If there is no interaction between the inputs, the effect of each input on the response is independent of values of other inputs. After conducting the sensitivity analysis, the insensitive inputs are set to some value and effects of the sensitive factors on the reservoir performance are investigated.

To determine the most influential factors, one can easily refer to the regression model (response surface model) and compare the regression coefficients. Usually, all input and response variables are standardized and the result of the regression model is then plotted on a Pareto chart. A variable is standardized by subtracting the variable value from its mean value and dividing the result by standard deviation of the variable [Santner et al., 2003].

3.5.5.1 Sensitivity analysis work flow

To perform a sensitivity analysis task, first a base reservoir model (base case) is created. In the base case, all sections of a reservoir model, including uncertain

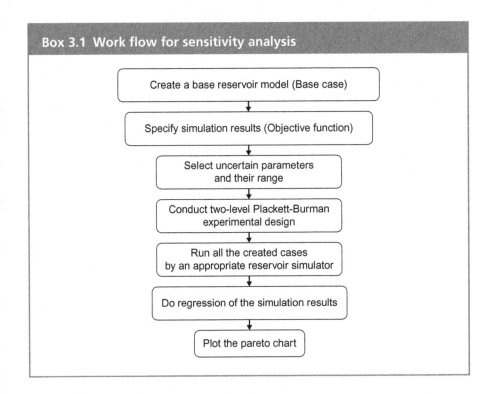

parameters, have values and the simulation results for studying the reservoir performance (objective functions) are specified, too. The next steps are to select parameters with uncertainty and to specify the range of values of these uncertainties (usually a wide range for each factor is specified). Then a suitable experimental design is chosen for screening the factors. Two-level Plackett-Burman design is the most commonly used design. By selecting an experimental design, several cases are created that should be run by a suitable reservoir simulation package. Regression models are next used to fit the specified simulation results. Drawing a Pareto chart is the final stage to assess the importance of the input uncertain factors. Any parameter with lower effect than the significant level should be discarded. Work flow for sensitivity analysis is shown in Box 3.1.

Case studies

4

4.1 Introduction

Experimental designs have been used in reservoir engineering since the 2000s [Cheong and Gupta, 2005; Corre et al., 2000; Friedmann et al., 2001; Khosravi et al., 2012; Peake et al., 2005; Portella et al., 2003; White and Royer, 2003; White et al., 2001]. The main purpose of using this method is to obtain the maximum amount of information with the lowest cost.

In this chapter, six different case studies of applications of experimental designs in reservoir studies are examined. In each case study, two-level Plackett-Burman experimental design is used for discovering the most influential parameters. After screening is done, a three-level experimental design is applied for predicting the reservoir performance. At the end of each case study, Monte Carlo simulations are performed to assess risk in the predicted reservoir performance. A list of the case studies is as follows:

- Case study 1: Ninth SPE comparative solution problem
- Case study 2: Undersaturated fractured reservoir in the Middle East
- Case study 3: PUNQ case
- Case study 4: SAGD in a heavy oil reservoir
- Case study 5: Barnett shale gas reservoir
- Case study 6: Miscible WAG injection case

The general steps for each case study are shown in Box 4.1.

4.2 Case study 1

The first case study describes a reservoir with a high degree of heterogeneity provided by geostatistically based permeability field data. Data for reservoir modeling are taken from the ninth SPE comparative solution project [Killough, 1995]. As geostatistics are used for generating reservoir permeability, it is expected that the generated permeability data suffer from uncertainty. Experimental designs are applied for uncertainty assessment in this case study.

4.2.1 Ninth SPE comparative solution problem

The reservoir of the ninth SPE comparative solution project was described by Killough in 1995. It was a dipping reservoir with a dipping angle of $10°$ in the x direction. The reservoir is divided into 15 layers with total thickness of 359 ft. Average porosity of the reservoir is 13%. Porosity and thickness of each layer are shown in Table 4.1. Layers 4, 6, 7 and 15 have porosities higher than 15%, and porosities of layers 1, 2 and 8 are less than 10%.

Experimental Design in Petroleum Reservoir Studies. DOI: http://dx.doi.org/10.1016/B978-0-12-803070-7.00004-1

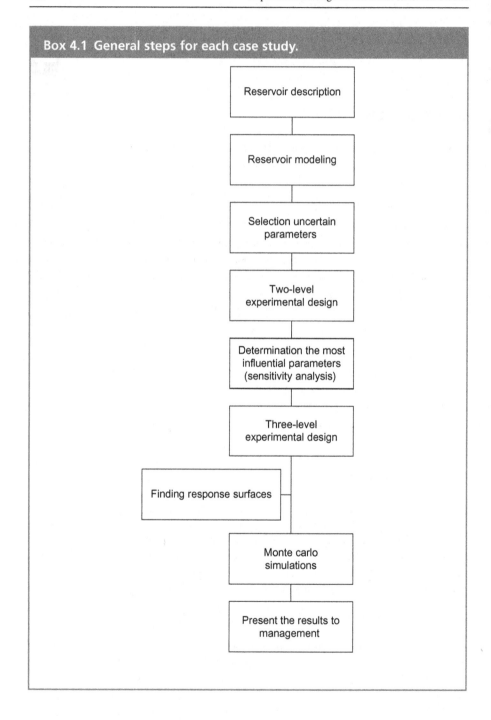

Box 4.1 General steps for each case study.

Table 4.1 Thickness and porosity of different layers, case study 1

Layer No.	Thickness (ft)	Porosity (fraction)
1	20	0.087
2	15	0.097
3	26	0.111
4	15	0.16
5	16	0.13
6	14	0.17
7	8	0.17
8	8	0.08
9	18	0.14
10	12	0.13
11	19	0.12
12	18	0.105
13	20	0.12
14	50	0.116
15	100	0.157

Table 4.2 PVT properties of case study 1

P (psi)	Rs (ft3/stb)	Bo (ft3/scf)	Z	μ_o (cP)	μ_g (cP)
14.7	0	1	0.9999	1.2	0.0125
400	165	1.012	0.8369	1.17	0.013
800	335	1.0255	0.837	1.14	0.0135
1200	500	1.038	0.8341	1.11	0.014
1600	665	1.051	0.8341	1.08	0.0145
2000	828	1.063	0.837	1.06	0.015
2400	985	1.075	0.8341	1.03	0.0155
2800	1130	1.087	0.8341	1	0.016
3200	1270	1.0985	0.8398	0.98	0.0165
3600	1390	1.11	0.8299	0.95	0.017
4000	1500	1.12	0.83	0.94	0.0175

Oil and gas PVT properties are given in Table 4.2. Initial reservoir pressure and temperature at a reference depth of 9035 ftss are 3600 psia and 100°F, respectively. The oil pressure gradient is 0.39 psia/ft at 3600 psia, density of the stock tank oil is 0.7206 gr/cc and the molecular weight of residual oil is 175. The stock tank water density is 1.0095 g/cc, water formation volume factor is 1.0034 RB/STB and its pressure gradient is 0.436 psi/ft.

The reservoir is initially undersaturated and its initial oil-water contact is estimated to be at 9950 ftss.

There are 25 production wells and one water injector. All production wells are completed in layers 2, 3 and 4. The water injection well is completed in layers

11,12,13,14 and 15. The maximum oil rate has been set to 1500 STB/day for 300 days. From the 300th day until the end of the first year of production, the oil rate has been dropped to 100 STB/day. Then the oil rate again was increased to 1500 STB/day and has remained unchanged until the end of the simulation (900 days). The minimum flowing bottom-hole pressure for all wells was set to 1000 psi at the depth of 9110 ftss. The maximum water injection rate was 5000 STB/day with a maximum bottom-hole pressure of 4000 psi at the depth of 9110 ftss. After 900 days of production, cumulative oil production was 18.24 MMSTB.

The purpose of this study is to build a reservoir model to match the cumulative oil production at the end of 900 days, and then to predict oil production after 5 years. To that purpose, the reservoir is discretized to 9000 (24 × 25 × 15) Cartesian grids. The sizes of the grids in the X and Y directions are 300 ft and the first cell (1,1,1) is located at a depth of 9000 ftss. Absolute permeability distribution is generated by geostatistics. The histogram of this property is shown in Figure 4.1. Relative permeabilities of oil-gas and oil-water are generated using the Corey correlation and are shown in Figure 4.2.

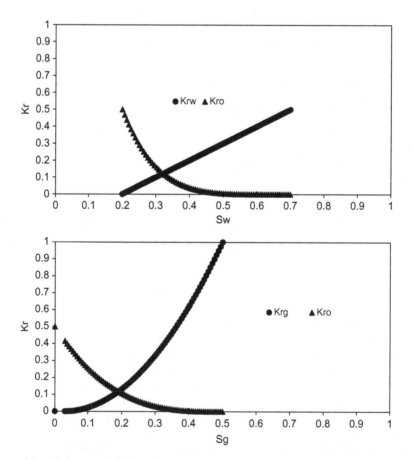

Figure 4.1 Relative permeability curves, case study 1.

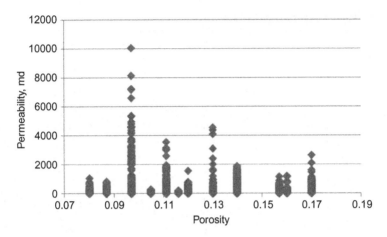

Figure 4.2 Permeability of different layers, case study 1.

Table **4.3** **Definition uncertain parameters, case study 1**

Parameter	Symbol	Low value	Mid value	High value
Krw at Swmax	KRW	0.2	0.6	1
Kro at Swc	KRO	0.2	0.6	1
Krg at Swc	KRG	0.2	0.6	1
Water-Oil contact (ft)	WOC	9900	9950	10000
Vertical permeability multiplier	Kz	0.1	1	10
Horizontal permeability multiplier	Kx	0.1	1	10

4.2.2 Uncertain parameters

Relative permeability data (as there is special core data), values of horizontal and vertical permeability (since they were generated by geostatistical techniques) and initial oil-water contact (as there is no water observation well) are considered as uncertain parameters. Using experimental design, effects of these parameters (maximum oil relative permeability, KRO, maximum water relative permeability, KRW, maximum gas relative permeability, KRG, vertical and horizontal permeability multipliers, Kz and Kx, and water-oil contact, WOC) on cumulative oil production are studied. Table 4.3 shows the uncertain parameters and their limits.

4.2.3 Experimental design

The first step is to screen the uncertain factors by a two-level experimental design. Here we apply Plackett-Burman design with a center as shown in Table 4.4. The analysis of this design is represented as a Pareto chart in Figure 4.3. This figure shows that all six factors are important parameters for cumulative oil production.

Table 4.4 **Plackett-Burman Design**

Run No.	Kx	Kz	WOC	KRG	KRO	KRW
RUN_001	0.1	10	10000	1.00	0.20	0.20
RUN_002	0.1	0.1	10000	0.20	1.00	1.00
RUN_003	10.0	0.1	9900	0.20	1.00	0.20
RUN_004	0.1	10	10000	0.20	1.00	1.00
RUN_005	0.1	0.1	9900	0.20	0.20	0.20
RUN_006	1.0	1	9950	0.60	0.60	0.60
RUN_007	10.0	10	10000	0.20	0.20	0.20
RUN_008	10.0	10	9900	0.20	0.20	1.00
RUN_009	10.0	0.1	10000	1.00	0.20	1.00
RUN_010	0.1	10	9900	1.00	1.00	0.20
RUN_011	0.1	0.1	9900	1.00	0.20	1.00
RUN_012	10.0	0.1	10000	1.00	1.00	0.20
RUN_013	10.0	10	9900	1.00	1.00	1.00

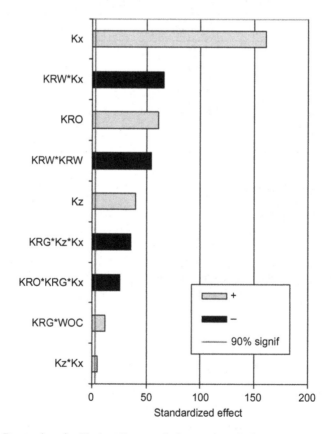

Figure 4.3 Pareto chart for Plackett-Burman design, case study 1.

To figure out the proper values of these six factors and in order to match the cumulative oil production at the end of 900 days of production, a three-level experimental design of Box-Behnken is then applied. The results are shown in Table 4.5, where the last two columns are cumulative oil production after 900 days and 5 years of production. This table shows that run 27 could match field oil production (18.2 MMSTB).

4.2.4 Response surfaces

In this stage of the study we try to find a relationship between cumulative oil production (FOPT) and the six selected parameters. Using nonlinear regression the following correlation is found:

$$\begin{aligned} FOPT = {} & b0 + b1 * Kx + b2 * KRO + b3 * KRG + b4 * Kz + b5 * WOC \\ & + b6 * KRG * Kx + b7 * Kx * Kx + b8 * KRO * KRO + b9 * KRG * KRG \\ & + b10 * KRO * Kz + b11 * Kz * Kz + b12 * KRO * KRG \end{aligned}$$

Regression coefficients (bi) of the response surface are shown in Table 4.6. Figures 4.4 and 4.5 represent response surfaces.

The above correlation is then used in a Monte Carlo simulation to find the cumulative distribution of FOPT. In the Monte Carlo simulation, distribution functions of each factor should be defined. In this case study, a normal distribution for each factor is selected. The result of 5000 realizations is shown in Figure 4.6, which represents that the reservoir cumulative oil production would probably be 26.16 MMSTB after 5 years.

4.3 Case study 2

Middle East carbonate reservoirs contain half of the world's oil [Middle East & Asia Reservoir Review, 1997]. Among the oil producer countries in the Middle East, Iran with 157.8 billion barrels proven reserve is the second country after Saudi Arabia [OPEC, 2014]. The main volume of the Iranian oil resources are accumulated in carbonate reservoirs, which are fractured and in general very complex reservoirs.

Generally, three components of a fractured reservoir are: 1) a fracture network; 2) filling materials within fractures; 3) matrix blocks between the fractures [Dietrich et al., 2005]. In Iranian carbonate fractured reservoirs, most of the hydrocarbon resides in the matrix, and fractures act as preferential flow paths. Oil flow conditions in the matrix are usually poor (due to low permeability of the matrix) and the time needed to produce the oil is longer than with sandstone reservoirs. Since there is almost no pressure drop in the fractures, water and gas can easily flow through fractures and this results in higher remaining oil saturation in the matrix block than the residual oil saturation in nonfractured reservoirs.

In order to characterize and to model a fractured reservoir, some properties such as fracture size, fracture distance, fracture density, fracture aperture, fracture

Table 4.5 Box-Behnken design and the results, case study 1

Run Number	KRW	KRO	KRG	WOC (ft)	Kz	Kx	FOPT at 900 days (MMSTB)	FOPT at 5 years (MMSTB)
1	1.00	0.60	1.00	9950	1	0.1	13.3	19.5
2	0.60	0.20	0.60	9950	10	0.1	7.578	11.65
3	0.60	0.60	0.20	10000	1	0.1	13.64	23.11
4	0.60	0.60	1.00	9900	1	0.1	12.79	18.24
5	0.60	0.60	0.20	9900	1	10.0	30.89	36.67
6	1.00	0.60	0.60	9900	0.1	1.0	24.21	29.29
7	0.60	0.60	0.60	9950	1	1.0	26.56	29.39
8	1.00	0.60	0.60	10000	10	1.0	24.67	24.91
9	0.20	0.20	0.60	9900	1	1.0	18.72	21.05
10	0.60	0.60	0.20	9900	1	0.1	12.81	20.88
11	0.60	0.60	0.20	10000	1	10.0	31.35	42.78
12	0.20	0.60	0.60	9900	10	1.0	21.29	21.35
13	0.20	1.00	0.60	9900	1	1.0	27.71	30
14	1.00	1.00	0.60	10000	1	1.0	29.35	34.91
15	1.00	1.00	0.60	9900	1	1.0	27.57	29.9
16	0.60	0.60	1.00	10000	1	0.1	13.7	20.69
17	1.00	0.60	0.60	10000	0.1	1.0	26.34	34.19
18	0.60	0.20	0.60	9950	10	10.0	20.84	20.84
19	0.60	1.00	0.20	9950	0.1	1.0	28.8	43.46
20	1.00	0.60	0.60	9900	10	1.0	20.91	20.94
21	0.20	0.60	0.20	9950	1	0.1	13.34	22.22
22	0.20	0.60	1.00	9950	1	0.1	13.33	19.45
23	0.60	0.60	0.60	9950	1	1.0	26.56	29.39
24	0.20	0.60	1.00	9950	1	10.0	28.9	28.93
25	0.20	1.00	0.60	10000	1	1.0	29.28	33.68
26	0.60	1.00	1.00	9950	0.1	1.0	27.71	32.86

27	1.00	0.20	0.60	9900	1	1.0	18.2	19.85
28	0.60	1.00	0.20	9950	10	1.0	28.74	34.05
29	0.20	0.60	0.60	10000	10	1.0	23.73	23.83
30	0.60	0.60	0.60	9950	1	1.0	26.56	29.39
31	1.00	0.60	0.20	9950	1	10.0	31.23	38.69
32	0.20	0.60	0.20	9950	1	10.0	30.99	41.39
33	0.60	0.60	1.00	9900	1	10.0	25.7	25.7
34	1.00	0.60	1.00	9950	1	10.0	27.16	27.17
35	0.60	0.20	0.20	9950	10	1.0	20.74	23.46
36	0.60	1.00	0.60	9950	10	10.0	29.41	29.41
37	0.60	1.00	0.60	9950	10	0.1	16.4	20.63
38	0.60	0.20	0.20	9950	0.1	1.0	17.9	26.33
39	0.60	0.20	0.60	9950	0.1	10.0	21.55	22.34
40	0.60	1.00	0.60	9950	0.1	10.0	31.34	37.42
41	1.00	0.60	0.20	9950	1	0.1	13.28	22.07
42	0.20	0.60	0.60	9900	0.1	1.0	24.58	30.05
43	0.60	0.20	1.00	9950	10	1.0	16.36	16.38
44	0.60	0.20	1.00	9950	0.1	1.0	16.68	20.95
45	0.60	0.20	0.60	9950	0.1	0.1	6.492	11.29
46	0.60	1.00	1.00	9950	10	1.0	22.89	22.9
47	0.60	1.00	0.60	9950	0.1	0.1	15.98	25.86
48	1.00	0.20	0.60	10000	1	1.0	20.78	24.3
49	0.20	0.20	0.60	10000	1	1.0	20.63	24.23
50	0.20	0.60	0.60	10000	0.1	1.0	26.41	34.02
51	0.60	0.60	1.00	10000	1	10.0	29.89	29.97

Table 4.6 **Coefficients of response surface, case study 1**

b0	29.00
b1	6.072
b2	5.518
b3	−3.850
b4	−3.238
b5	1.945
b6	−2.335
b7	−3.374
b8	−1.662
b9	1.742
b10	−1.501
b11	−1.580
b12	−1.161

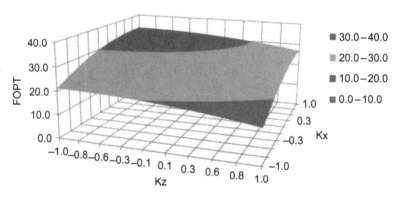

Figure 4.4 Effect of vertical and horizontal permeabilities on field oil production, case study 1.

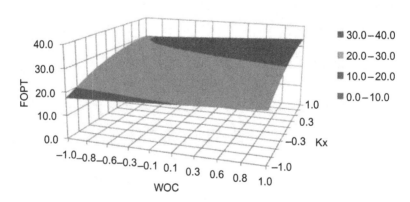

Figure 4.5 Effect of horizontal permeability and water oil contact on field oil production, case study 1.

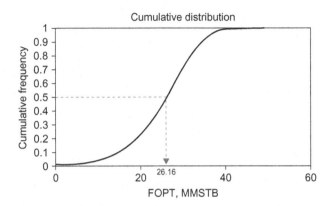

Figure 4.6 Cumulative distribution of field oil production, case study 1.

orientation, and matrix block size should be identified. However, due to geometrical complexity and the large number of fracture/matrix and matrix/matrix exchange processes involved, modeling of fluid flow through naturally fractured reservoirs leads to several simplifications. The three most common models are the double-porous and single-permeable (DPSP) model, the double-porous and double-permeable (DPDP) model, and the single-porous and single-permeable (SPSP) model [Dietrich et al., 2005].

The second case study explains an undersaturated fractured carbonate reservoir.

4.3.1   Undersaturated fractured reservoir in the middle east

The reservoir is an anticline carbonate reservoir where diagenesis processes caused dissolution, dolomitization and cementation. Geological studies show that there are fractures and connected vugs through the reservoir. Based on lithology, the formation is divided into seven zones. The zones contain dolomitized limestone, anhydrides and shale. Sampled cores were transferred to the core lab and more than 360 plugs were cut from them, in order to conduct routine and special core tests. A Lorenz plot shows that the reservoir has a high degree of heterogeneity (Figure 4.7). As the reservoir is highly heterogeneous, we expect more than one rock-type. A cross plot of permeability-porosity is shown in Figure 4.8. Based on this plot, the reservoir is divided into four rock-types: rocks with porosities of 5% to 8% are classified in the first rock-type; in the second rock-type, porosities are less than 12%; rocks that have porosities of less than 16% are in the third rock-type; and in the fourth rock-type, porosities are greater than 16%. As the reservoir is highly heterogeneous, we expect more than one rock-type. Figure 4.9 shows a plot of normalized Leverett J-function versus normalized saturation. Relative permeability and capillary pressure data for those four rock-types are shown in Figures 4.10 to 4.13.

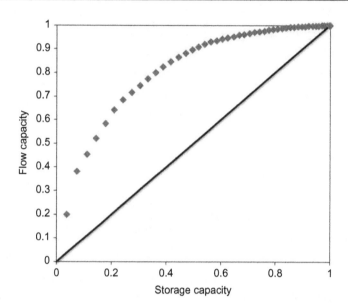

Figure 4.7 Lorenz plot, case study 2.

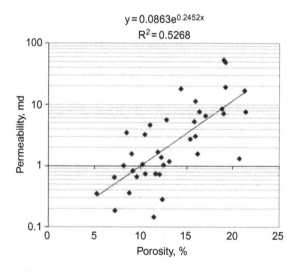

Figure 4.8 Permeability-porosity cross plot, case study 2.

Seven production wells were drilled vertically and completed as open-hole completion in the reservoir. There is a 14-year history of oil production. There is a well testing data for a well of this reservoir.

The analysis of the well testing data is done using the FAST WellTest™ software (IHS, 2014). Results of the analysis are shown in Figures 4.14 to 4.16. Initial reservoir pressure and temperature were 4100 psia and 220°F, respectively, at a reference depth

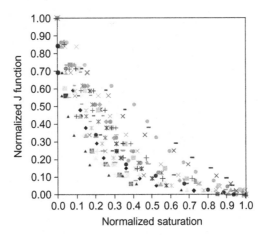

Figure 4.9 Leverett J-function, case study 2.

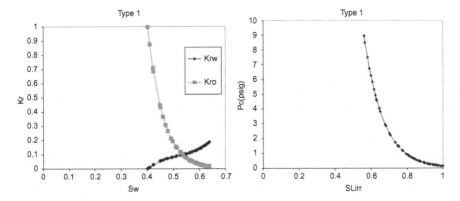

Figure 4.10 Relative permeability and capillary pressure curves for rock type 1, case study 2.

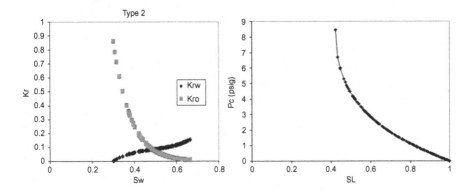

Figure 4.11 Relative permeability and capillary pressure curves for rock type 2, case study 2.

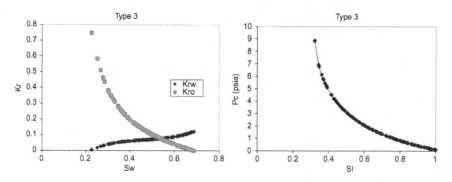

Figure 4.12 Relative permeability and capillary pressure curves for rock type 3, case study 2.

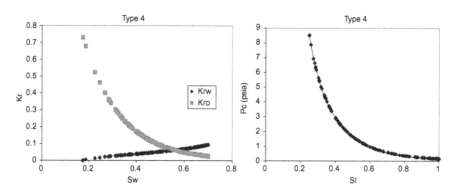

Figure 4.13 Relative permeability and capillary pressure curves for rock type 4, case study 2.

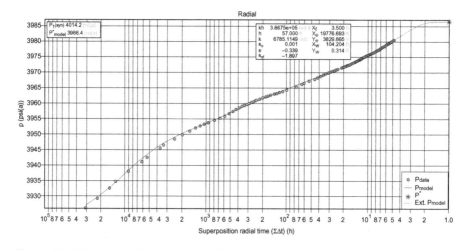

Figure 4.14 Well test analysis, case study 2.

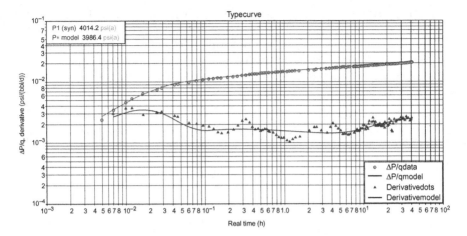

Figure 4.15 Well test analysis, case study 2.

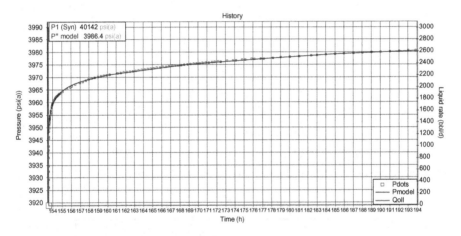

Figure 4.16 Well test analysis, case study 2.

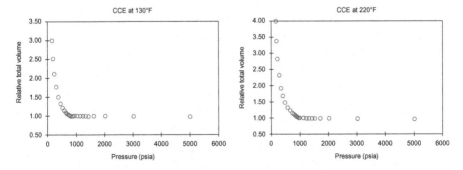

Figure 4.17 Results of constant composition expansion experiments, case study 2.

of 7800 ftss. Preliminary fluid analysis on a bottom-hole sample showed a bubble point pressure of 990 psia. So, the reservoir fluid is an undersaturated oil reservoir. Complete PVT tests (two constant composition expansion tests and one differential liberation test) were conducted on sampled fluid. Figures 4.17 and 4.18 represent the results of PVT tests. Compositions of hydrocarbons are shown in Table 4.7.

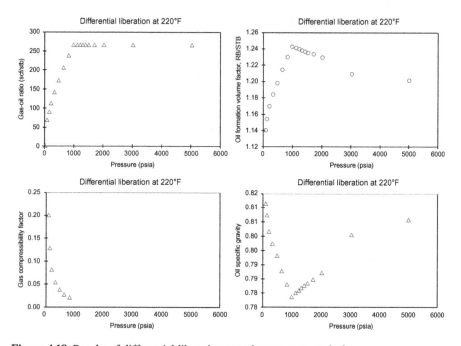

Figure 4.18 Results of differential liberation experiments, case study 2.

Table 4.7 **Composition of fluid, case study 2**

Component	Mole %
C1	16.4
C2	5.79
C3	7.05
IC4	1.22
NC4	4.13
IC5	2.06
NC5	2.44
FC6	4.61
C7+	56.3
C7+ specification	
Specific gravity Molecular weight	0.916 280

To model phase behavior, the Peng-Robinson equation of state [Peng & Robinson, 1976, 1977] is applied:

$$p = \frac{RT}{V-b} - \frac{a\alpha}{V^2 + 2bV - b^2}$$

$$a = 0.457235 \frac{R^2 T_C^2}{p_C} = \Omega_a \frac{R^2 T_C^2}{p_C}$$

$$b = 0.077796 \frac{RT_C}{p_C} = \Omega_b \frac{RT_C}{p_C} \tag{4.1}$$

$$\alpha = (1 + \kappa(1 - T_r^{0.5}))^2$$

$$\kappa = 0.37464 + 1.54226\omega - 0.26992\omega^2$$

Three parameters (a, b and ω) of this equation of state should be adjusted to match the results of the equation of state with PVT tests. For that purpose, the heavy-end fraction (C7+) is split into several single carbon numbers and then lumped into six groups, as shown in Table 4.8.

As Coats and Smart (1986) stated, a cubic equation of state does not accurately predict laboratory data of oil/gas mixtures without the tuning of the EOS parameters. Thus, to figure out the proper values of parameters a, b and ω, all experimental points (data) are used in tuning the equation of state. The objective of regression is to minimize the difference between experimental data and predicted results of the equation of state:

$$F = \sum w_i \left(\frac{y_{EOS} - y_{Exp}}{y_{Exp}} \right)^2 \tag{4.2}$$

Table 4.8 Composition of fluid after splitting and lumping, case study 2

Component	Mole %
C1	16.40
C2	5.79
C3	7.05
IC4	1.22
NC4	4.13
IC5	2.06
NC5	2.44
FC6	4.61
C07−C12	20.12
C13−C17	11.15
C18−C23	8.94
C24−C29	5.75
C30	0.73
C31+	9.60

where w_i is a weight factor and indices *EOS* and *Exp* refer to the equation of state and experiment, and y_i represents experimental data.

The regression is done using the adaptive least-squares algorithm of Dennis, Gay & Welsch [1981] modified by Chen and Stadtherr [1981]. Selection of regression parameters is done dynamically from a larger set of variables (variables with larger dF/dx, where x is the selected parameter scaled by using the upper bound $x_{j,max}$ and lower bound $x_{j,min}$ of the corresponding parameter). Here, the WINPROP module of Computer Modeling Group (CMG) Ltd. [2011] is applied: critical pressure, critical temperature, critical volume, acentric factor, Ω_a and the power of the Chueh and Prausnitz (1967) equation for binary interaction coefficients are selected for FC6 to C31+:

$$d_{ij} = 1 - \left[\frac{2V_{ci}^{1/6} V_{cj}^{1/6}}{V_{ci}^{1/3} + V_{cj}^{1/3}} \right]^{\theta} \tag{4.3}$$

Results of the regression are represented in Figures 4.19 and 4.20. Table 4.9 shows the new values for the selected regression parameters.

In order to build the reservoir model, the reservoir is discretized into 310,464 ($84 \times 42 \times 88$) grid cells and the model is considered as a dual porosity reservoir (so half of the grids stand for fractures). In a Warren and Root (1963) dual porosity model, the fluid flow through the reservoir takes place only in the fracture network with the matrix blocks acting as sources. This model allows each block to have up to two porosity systems, one called the matrix porosity and the other called the fracture porosity. Each medium may have its own porosity value and permeability, as well as other distinctive properties (Figure 4.21). In the Warren and Root model, inter-block flows are governed by the fracture properties; the matrix-fracture flow is proportional to a transmissibility that is calculated by using the matrix-fracture shape factor:

Transmissibility = (matrix permeability)(matrix cell bulk volume)(shape factor)

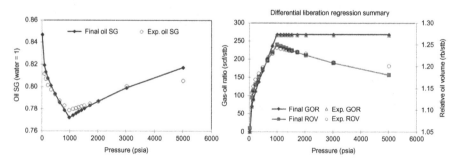

Figure 4.19 Tuning P-R EOS with differential liberation experiments, case study 2.

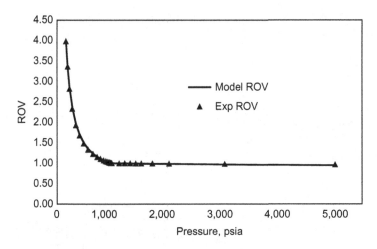

Figure 4.20 Tuning P-R EOS with constant composition expansion experiments, case study 2.

Table 4.9 **Properties of heavy components after tuning EOS, case study 2**

Component	P_C(atm)	T_C (K)	ω	MW	Ω_a
FC6	29.805	560.77	0.31209	68.8	0.37382
C07–C12	23.41202	674.54	0.4487	102.93	0.3715
C13–C17	17.85516	680.32	0.6355	248.1	0.51298
C18–C23	15.450622	848.27	0.8089	339.54	0.34198
C24–C29	14.66021	903.74	0.81768	293.68	0.34198
C30	10.87415	905.67	1.11038	421.0573	0.36579
C31+	10.72163	842.2	1.12298	686.5867	0.36579

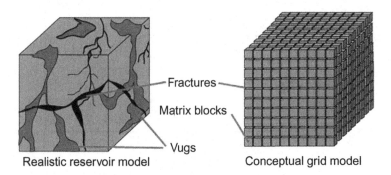

Realistic reservoir model   Conceptual grid model

Figure 4.21 Dual porosity model [Warren and Root, 1963].

Kazemi [1976] has proposed the following equation for shape factor:

$$\sigma = 4\left(\frac{1}{l_x^2} + \frac{1}{l_y^2} + \frac{1}{l_z^2}\right)$$

(4.4)

where σ is shape factor and l_x, l_y and l_z are typical x, y and z dimensions of the blocks of material making up the matrix volume.

Figure 4.22 shows a 3 D view of the reservoir.

Normally, relative permeability of fluids in the fractures is assumed linearly (Figure 4.23) and the capillary pressure is set to zero.

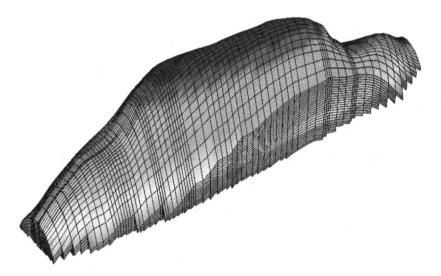

Figure 4.22 Three-dimensional view of the reservoir, case study 2.

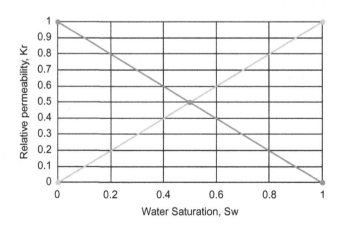

Figure 4.23 Relative permeability for fractures, case study 2.

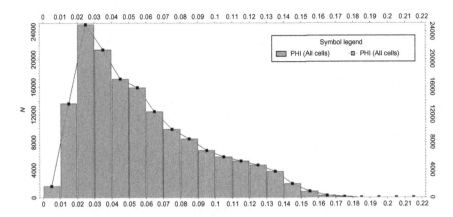

Figure 4.24 Histogram of porosity, case study 2.

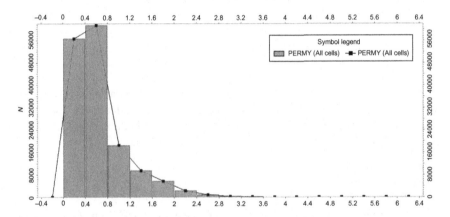

Figure 4.25 Histogram of permeability, case study 2.

Histograms of matrix porosity and permeability (that were generated by Gaussian geostatistical technique) are shown in Figures 4.24 and 4.25, respectively. Based on core appearance and previous experience fracture porosity and permeability are set to 0.002 and 600 md, respectively.

Through using the material balance technique, the presence of an aquifer could be easily studied. To do that, one can write the material balance equation for an undersaturated reservoir:

$$F = N(E_0 + E_{f,w}) + (W_i + W_e)B_W \tag{4.5}$$

$$F = N_P(B_o) + W_P B_w \tag{4.6}$$

Expansion of oil and solution gas can be written as:

$$E_0 = (B_o - B_{oi}) \tag{4.7}$$

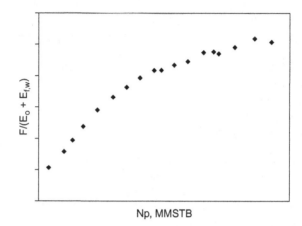

Figure 4.26 Results of linear material balance analysis, case study 2.

$$\Delta E_{f,w} = -B_{oi}\left(\frac{C_r + C_w S_{wi}}{1 - S_{wi}}\right)\Delta P \tag{4.8}$$

A plot of $F/(E_o + E_{f,w})$ versus N_p could give information about aquifers [Dake, 1978]. Figure 4.26 contains that plot, depicting that there is a strong aquifer in this reservoir.

In order to define the aquifer in the reservoir model, a bottom-edge Carter Tracy analytical aquifer is used with the following initial properties:

permeability = 5 md
porosity = 0.1
thickness = 500 ft.

As stated, there are seven production wells in the reservoir. All wells are open-holed and vertical. In the simulation, the rate of each well is set to observed-well rate (rate control mode) and wells bottom-hole pressures are calculated and compared to the observed pressure data.

4.3.2 Uncertainty parameters

In this study, aquifer properties, fracture porosity, block height and water-oil level are selected as uncertain parameters. Table 4.10 shows the uncertain parameters and their extremes.

Plackett-Burman with a center point design is used in order to study the effect of each uncertain parameter on reservoir performance, as presented in Table 4.11.

These 13 cases were run using the ECLIPSE Black-Oil simulator [Schlumberger, 2011]. As we enforce the simulator to catch the rate, in all 13 cases oil rates match with the observed data, as shown in Figure 4.27. But average reservoir pressure is different for different values of the uncertain parameters (see Figure 4.28).

Figures 4.29 to 4.34 show Pareto charts for recognizing the most influential factors on well-block pressures of six wells. It's clear that from the six chosen uncertain

Table **4.10** **Uncertain parameters and their extremes, case study 2**

Parameter	Description	Unit	Low level	Middle level	High level
AQPERM	Aquifer permeability	md	1	5.5	10
AQPOR	Aquifer porosity	fraction	0.07	0.1	0.13
AQTHICK	Aquifer thickness	ft	400	500	600
SIGMA	Shape factor	$1/ft^2$	0.020	0.040	0.060
FracPor	Fracture porosity	fraction	0.001	0.002	0.003
WOC	Oil-water level	ft	8400	8425	8450

Table **4.11** **Plackett-Burman with a center point design**

Case Number	AQPERM	AQPOR	AQTHICK	SIGMA	FracPor	WOC
1	1.00	0.07	600.000	0.06	0.003	8400
2	10.00	0.13	400.000	0.06	0.001	8400
3	1.00	0.13	400.000	0.02	0.001	8450
4	10.00	0.13	400.00	0.06	0.003	8400
5	1.00	0.007	400.00	0.02	0.001	8400
6	5.50	0.10	500.000	0.04	0.002	8425
7	1.00	0.07	400.00	0.06	0.003	8450
8	10.00	0.07	400.000	0.02	0.003	8450
9	10.00	0.07	600.00	0.06	0.001	8450
10	1.00	0.13	600.000	0.02	0.003	8400
11	10.00	0.07	600.00	0.02	0.001	8400
12	1.00	0.13	600.00	0.06	0.001	8450
13	10.00	0.13	600.00	0.02	0.003	8450

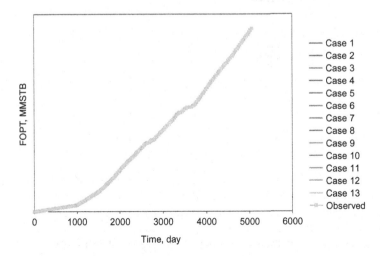

Figure 4.27 Field oil production total for all cases, case study 2.

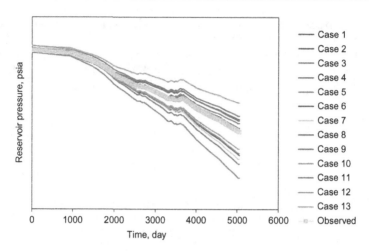

Figure 4.28 Reservoir pressure versus time for all cases, case study 2.

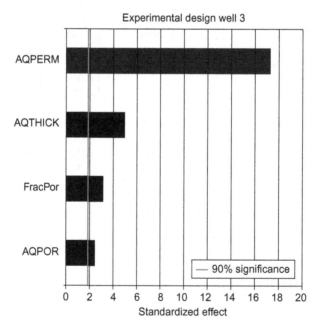

Figure 4.29 Pareto chart for well pressure of well 3, case study 2.

factors, only four of them (AQPERM, AQTHICK, FracPor and AQPOR) have impacts on pressure, while the other two factors, SIGMA and WOC, are insignificant.

The next stage is to make response surfaces. Using linear regression, response surfaces for those six wells (there is no available data for one of the wells) are generated and are shown in Tables 4.12 to 4.17 and Figures 4.35 to 4.40.

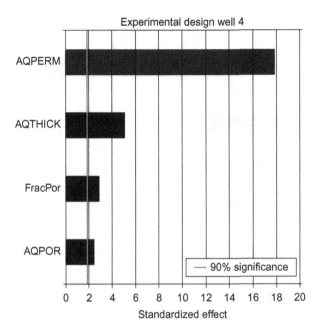

Figure 4.30 Pareto chart for well pressure of well 4, case study 2.

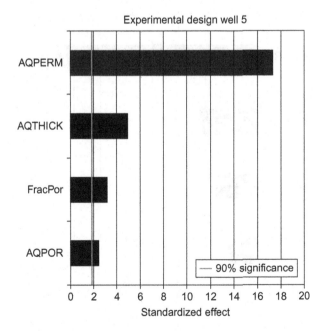

Figure 4.31 Pareto chart for well pressure of well 5, case study 2.

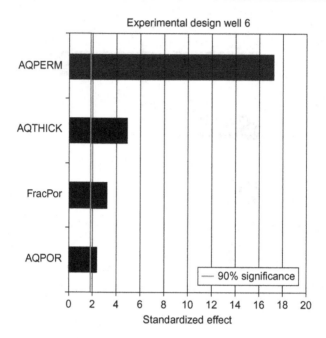

Figure 4.32 Pareto chart for well pressure of well 6, case study 2.

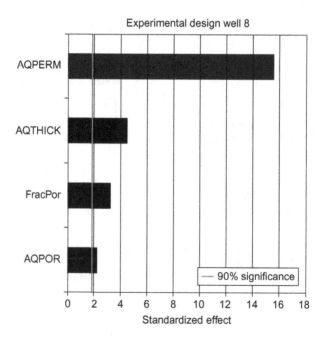

Figure 4.33 Pareto chart for well pressure of well 8, case study 2.

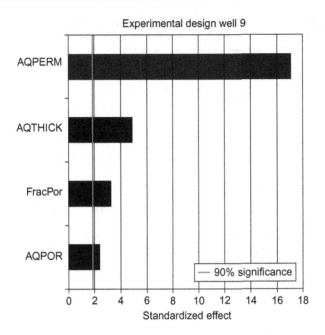

Figure 4.34 Pareto chart for well pressure of well 9, case study 2.

Table 4.12 Response surface for well pressure of well 3, case study 2

WPR3 = b0 + b1*AQPERM + b2*AQTHICK + b3*FracPor + b4*AQPOR	
b0	3925.8
b1	50.38
b2	14.35
b3	9.089
b4	7.095

Table 4.13 Response surface for well pressure of well 4, case study 2

WPR4 WPR3 = b0 + b1*AQPERM + b2*AQTHICK + b3*FracPor + b4*AQPOR	
b0	3891.9
b1	46.92
b2	13.35
b3	7.484
b4	6.424

Table 4.14 **Response surface for well pressure of well 5, case study 2**

WPR5 = b0 + b1*AQPERM + b2*AQTHICK + b3*FracPor + b4*AQPOR	
b0	3908.0
b1	49.63
b2	14.17
b3	9.089
b4	6.984

Table 4.15 **Response surface for well pressure of well 6, case study 2**

WPR6 = b0 + b1*AQPERM + b2*AQTHICK + b3*FracPor + b4*AQPOR	
b0	3900.8
b1	49.22
b2	14.06
b3	9.188
b4	6.910

Table 4.16 **Response surface for well pressure of well 8, case study 2**

WPR8 = b0 + b1*AQPERM + b2*AQTHICK + b3*FracPor + b4*AQPOR	
b0	3896.4
b1	47.89
b2	13.85
b3	10.07
b4	6.875

Table 4.17 **Response surface for well pressure of well 9, case study 2**

WPR9 = b0 + b1*AQPERM + b2*AQTHICK + b3*FracPor + b4*AQPOR	
b0	3898.7
b1	48.41
b2	13.90
b3	9.276
b4	6.805

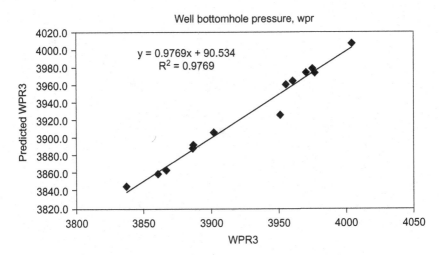

Figure 4.35 Regression result for well pressure of well 3, case study 2.

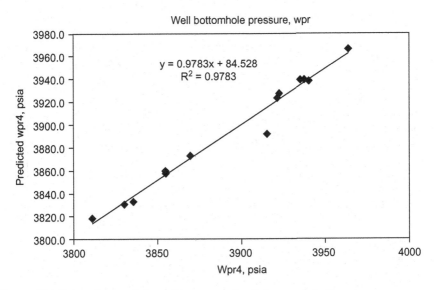

Figure 4.36 Regression result for well pressure of well 4, case study 2.

Using created response surfaces, proper distribution of uncertain factors (Table 4.18) and Monte Carlo simulation, one can generate cumulative distribution of well-block pressures for the wells, shown in Figures 4.41 to 4.46. Table 4.19 compares well block pressures with P50 (most likely values) of Monte Carlo. Based

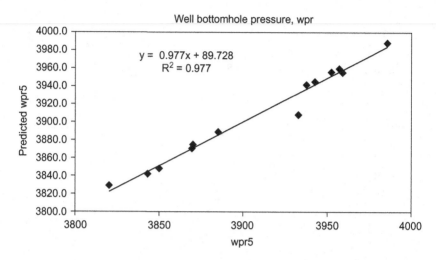

Figure 4.37 Regression result for well pressure of well 5, case study 2.

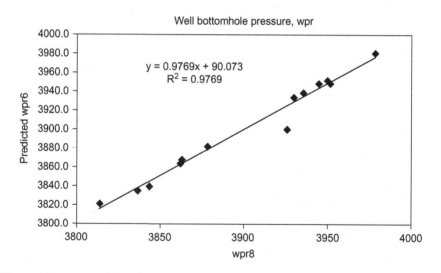

Figure 4.38 Regression result for well pressure of well 6, case study 2.

on the Monte Carlo simulation, values of the uncertain parameters for having a matched case are as follows:

AQPERM	5.50 md
AQPOR	0.09
AQTHICK	539.947 ft
FracPor	0.002

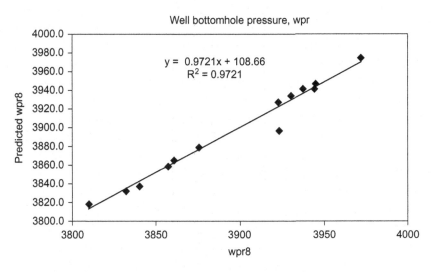

Figure 4.39 Regression result for well pressure of well 8, case study 2.

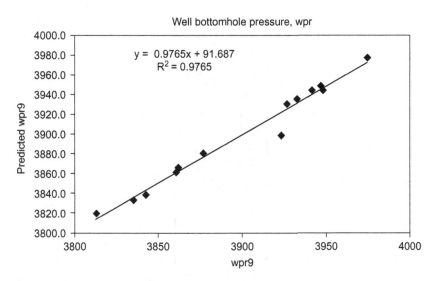

Figure 4.40 Regression result for well pressure of well 9, case study 2.

The average reservoir pressure of the matched case is then compared with the observed pressure in Figure 4.47.

The matched case is then used to predict reservoir performance for five years, Figure 4.48. So if there is no new drilled well in the reservoir, the reservoir is expected to produce more than 90 MMSTB oil and 20 bcf gas after 19 years production.

Table 4.18 Information for distribution functions, case study 2

Parameter name	Parameter type	Distribution	Definition values		
AQPERM	Cont	Normal	Mean = 5.5	Stdev = 1.2	Min = 1	Max = 10
AQPOR	Cont	Normal	Mean = 0.1	Stdev = 0.009	Min = 0.07	Max = 0.13
AQTHICK	Cont	Triangle	Low = 400	Mid = 500	High = 600	–
FracPor	Cont	Normal	Mean = 0.002	Stdev = 0.0003	Min = 0.001	Max = 0.003

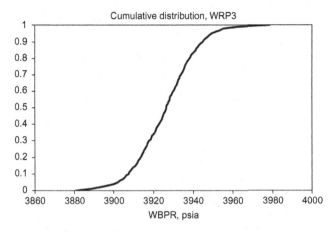

Figure 4.41 Monte Carlo results for well pressure of well 3, case study 2.

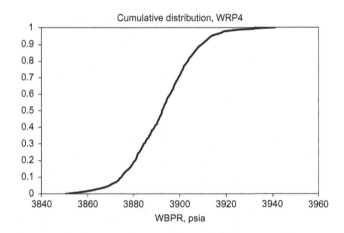

Figure 4.42 Monte Carlo results for well pressure of well 4, case study 2.

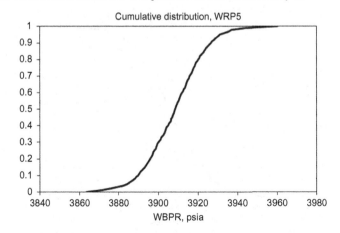

Figure 4.43 Monte Carlo results for well pressure of well 5, case study 2.

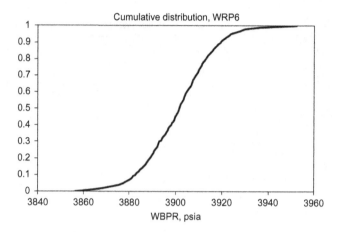

Figure 4.44 Monte Carlo results for well pressure of well 6, case study 2.

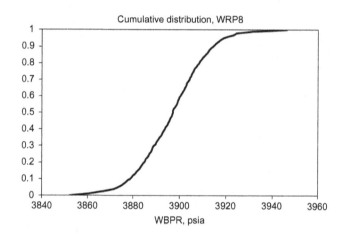

Figure 4.45 Monte Carlo results for well pressure of well 8, case study 2.

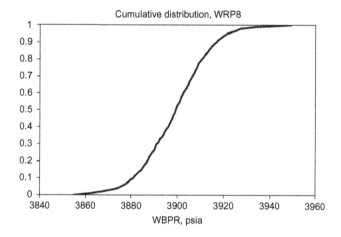

Figure 4.46 Monte Carlo results for well pressure of well 9, case study 2.

Table 4.19 **Monte Carlo results compared with observed data, case study 2**

Well Name	Observed	MC, P50	Error, %
Well3	3944.767	3926	0.475747
Well4	3916.998	3892	0.638193
Well5	3920.514	3908	0.31919
Well6	3919.21	3901	0.464645
Well8	3926.49	3896	0.776521
Well9	3909.559	3899	0.270089

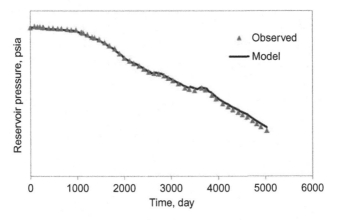

Figure 4.47 Model results compared with observed data, case study 2.

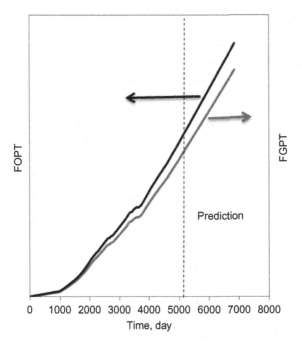

Figure 4.48 Prediction of total field oil production, case study 2.

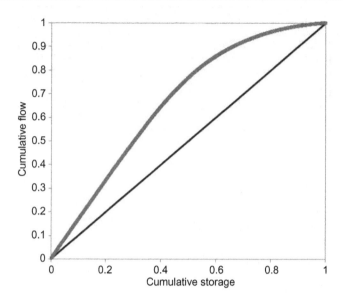

Figure 4.49 Lorenz plot, case study 3.

4.4   Case study 3

4.4.1   PUNQ case

The PUNQ (Production forecasting with UNcertainty Quantification) case is a synthetic reservoir model taken from a reservoir engineering study on a real North Sea reservoir operated by Elf Exploration Production [Zhang, 2003]. In PUNQ the problem was set up as a test case to study uncertainties in reservoir performance predictions.

This field is bounded to the east and south by a fault, and links to the north and west to a fairly strong aquifer. A small gas cap is located in the center of the dome-shaped structure. PVT properties are presented in Table 4.20. The field contains six production wells located around the gas-oil contact. The objective of this study is to find the cumulative oil production after 16.5 years production.

The reservoir model contains 2660 ($19 \times 28 \times 5$) corner point grid blocks, of which 1,761 blocks are active. Porosity and permeability are generated using the geological/geostatistical model. A Lorenz plot is shown in Figure 4.49 indicating the heterogeneity of the reservoir. A porosity-permeability cross plot is presented in Figure 4.50 and histograms of porosity and permeability for the reservoir are shown in Figures 4.51 and 4.52. Relative permeability curves are correlated by Corey and shown in Figure 4.53. Specifications and locations of the six production wells are depicted in Figure 4.54 and Table 4.21. The reservoir model is then simulated using the ECLIPSE Black-Oil simulator.

Table 4.20 **PVT properties for oil and gas, case study 3**

GOR (sm3/sm)	Pressure (bar)	Bo (rm3/sm3)	Oil viscosity (cP)
11.46	40	1.064	4.338
17.89	60	1.078	3.878
24.32	80	1.092	3.467
30.76	100	1.106	3.1
37.19	120	1.12	2.771
43.62	140	1.134	2.478
46.84	150	1.141	2.343
50.05	160	1.148	2.215
53.27	170	1.155	2.095
56.49	180	1.162	1.981
59.7	190	1.169	1.873
62.92	200	1.176	1.771
66.13	210	1.183	1.674
69.35	220	1.19	1.583
72.57	230	1.197	1.497
74	234.46 (Pb)	1.2	1.46
74*	250*	1.198*	1.541*
74*	300*	1.194*	1.787*
80	245	1.22	1.4
80*	300*	1.215*	1.7*

Pressure (bar)	Bg (rm3/sm3)	Gas viscosity (cP)
40	0.02908	0.0088
60	0.01886	0.0092
80	0.01387	0.0096
100	0.01093	0.01
120	0.00899	0.0104
140	0.00763	0.0109
150	0.00709	0.0111
160	0.00662	0.0114
170	0.0062	0.0116
180	0.00583	0.0119
190	0.00551	0.0121
200	0.00521	0.0124
210	0.00495	0.0126
220	0.00471	0.0129
230	0.00449	0.0132
234.46	0.0044	0.0133

*Undersaturated data.

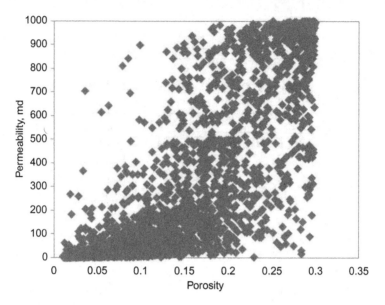

Figure 4.50 Permeability-porosity cross plot, case study 3.

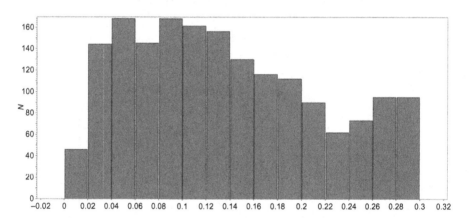

Figure 4.51 Histogram of porosity, case study 3.

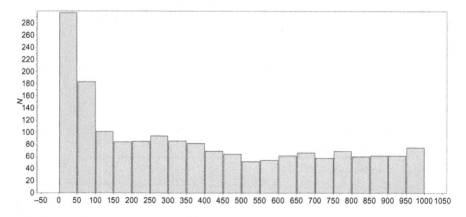

Figure 4.52 Histogram of permeability, case study 3.

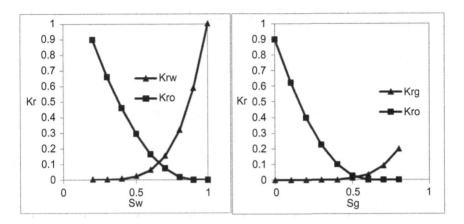

Figure 4.53 Relative permeability curves, case study 3.

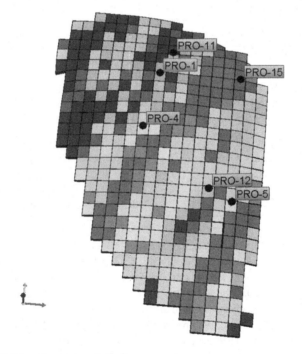

Figure 4.54 Well locations, case study 3.

4.4.2 Uncertain parameters

As the porosity and permeability of the reservoir are generated by geostatistical model, these parameters (porosity and permeability multipliers are defined for changing the porosity and permeability data) for all five layers are considered as uncertain parameters (10 parameters). A two-level Plackett-Burman with a center point design of

Table 4.21 Well specification and completion data (schedule section of ECLIPSE black-oil simulator)

WELL SPECIFICATION DATA				
WELSPECS (Well specification)				
Well name	**Group name**	**Well cell (i j)**	**Reference depth**	**Well phase**
'PRO-1'	'G1'	10 22	2362.2	'OIL'/
'PRO-4'	'G1'	9 17	2373.0	'OIL'/
'PRO-5'	'G1'	17 11	2381.7	'OIL'/
'PRO-11'	'G1'	11 24	2386.0	'OIL'/
'PRO-12'	'G1'	15 12	2380.5	'OIL'/
'PRO-15'	'G1'	17 22	2381.0	'OIL'/

WELL COMPLETION DATA			
COMPDAT (Well completion data)			
Well name	**Well cell (i j k1 k2)**	**Well status**	**Well diameter**
'PRO-1'	10 22 5 5	'OPEN'	0.15/
'PRO-1'	10 22 4 4	'OPEN'	0.15/
'PRO-4'	9 17 5 5	'OPEN'	0.15/
'PRO-4'	9 17 4 4	'OPEN'	0.15/
'PRO-5'	17 11 4 4	'OPEN'	0.15/
'PRO-5'	17 11 3 3	'OPEN'	0.15/
'PRO-11'	11 24 4 4	'OPEN'	0.15/
'PRO-11'	11 24 3 3	'OPEN'	0.15/
'PRO-12'	15 12 5 5	'OPEN'	0.15/
'PRO-12'	15 12 4 4	'OPEN'	0.15/
'PRO-15'	17 22 4 4	'OPEN'	0.15/

experiments is then applied in order to clarify the effect of the uncertain parameters on cumulative oil and gas production (Table 4.22). The last two columns of this table represent model results for the cumulative oil production (FOPT) and cumulative gas production (FGPT). Pareto charts of these two responses are depicted in Figure 4.55. The Pareto charts show that permeabilities of layers 2 and 3 (Kx2 and Kx3) have no effect on cumulative production of oil and gas (FOPT and FGPT).

In the next step, three-level fractional factorial design with IV resolution is used for these eight effective factors (see Table 4.23). Seventeen cases should be studied.

The last two columns of Table 4.23 show the results for FOPT and FGPT after 16.5 years production.

Response surfaces are then created by linear and nonlinear regression for FOPT and FGPT:

$$FOPT = b0 + b1 * Kx1 + b2 * POR5 + b3 * POR4 + b4 * Kx4 + b5 * Kx5 + b6 * POR1 + b7 * POR2 + b8 * POR3$$

$$(4.9)$$

Table 4.22 Plackett-Burman design for ten uncertain factors, case study 3

Run number	Kx1	Kx2	Kx3	Kx4	Kx5	POR1	POR2	POR3	POR4	POR5	FOPT (1000 m³)	FGPT (10⁶ m³)
Run1	0.50	0.50	1.50	1.50	1.50	0.50	1.50	1.50	0.50	1.50	4082	358.2
Run2	1.50	1.50	0.50	1.50	0.50	0.50	0.50	1.50	1.50	1.50	3798	353.5
Run3	0.50	1.50	0.50	0.50	0.50	1.50	1.50	1.50	0.50	1.50	3679	324
Run4	1.50	1.50	1.50	1.50	1.50	0.50	1.50	0.50	0.50	0.50	3223	272.4
Run5	0.50	0.50	0.50	0.50	0.50	0.50	0.50	0.50	0.50	0.50	2961	272.4
Run6	1.00	1.00	1.00	1.00	1.00	1.00	1.00	1.00	1.00	1.00	3787	336.1
Run7	0.50	0.50	1.50	1.50	1.50	1.50	0.50	1.50	1.50	0.50	3682	300
Run8	1.50	0.50	1.50	0.50	1.50	1.50	1.50	0.50	1.50	1.50	3686	317.5
Run9	1.50	0.50	0.50	1.50	0.50	1.50	0.50	0.50	0.50	1.50	3715	325
Run10	0.50	1.50	1.50	0.50	1.50	0.50	0.50	0.50	1.50	1.50	3908	348.2
Run11	1.50	0.50	0.50	0.50	0.50	0.50	1.50	1.50	1.50	0.50	3679	319.5
Run12	0.50	1.50	0.50	1.50	0.50	1.50	1.50	0.50	1.50	0.50	3771	321.4
Run13	1.50	1.50	1.50	0.50	1.50	1.50	0.50	1.50	0.50	0.50	3600	299.4

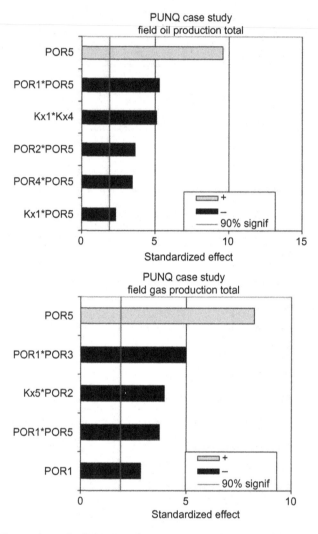

Figure 4.55 Pareto charts for field oil and gas production, case study 3.

$$\begin{aligned} FGPT = b0 + b1 * POR5 + b2 * Kx4 + b3 * POR4 \\ + b4 * POR3 + b5 * Kx1 * POR4 \end{aligned} \quad (4.10)$$

Tables 4.24 and 4.25 represent the coefficients of the created surface responses.

Using these response surfaces and Monte Carlo simulations, most likely values of cumulative oil and gas production are calculated and shown in Figure 4.56.

Table 4.23 Three-level fractional factorial design, case study 3

Run Number	Kx1	POR5	POR4	Kx4	Kx5	POR1	POR2	POR3	FOPT (1000 m³)	FGPT (10⁶ m³)
Run1	0.50	1.50	1.50	1.50	0.50	1.50	0.50	0.50	3873	348.3
Run2	1.50	1.50	0.50	1.50	0.50	0.50	0.50	1.50	3811	348.7
Run3	0.50	0.50	1.50	1.50	1.50	0.50	0.50	1.50	3736	314.8
Run4	0.50	1.50	1.50	0.50	0.50	0.50	1.50	1.50	3838	342.5
Run5	0.50	1.50	0.50	1.50	0.50	1.50	0.50	1.50	3926	338.2
Run6	0.50	0.50	0.50	1.50	1.50	1.50	1.50	1.50	3669	314.1
Run7	1.50	0.50	0.50	0.50	1.50	0.50	1.50	1.50	3443	284.9
Run8	0.50	1.50	0.50	1.50	1.50	0.50	1.50	0.50	3824	346.4
Run9	1.50	1.50	1.50	0.50	1.50	0.50	0.50	0.50	3763	339.4
Run10	1.50	0.50	0.50	1.50	1.50	1.50	0.50	0.50	3351	282
Run11	1.50	1.50	1.50	1.50	1.50	1.50	1.50	1.50	4031	347
Run12	0.50	0.50	0.50	0.50	0.50	0.50	0.50	0.50	3086	279.1
Run13	1.50	0.50	1.50	0.50	0.50	1.50	0.50	1.50	3634	315.1
Run14	0.50	0.50	1.50	0.50	1.50	1.50	1.50	0.50	3490	283.5
Run15	1.50	1.50	0.50	0.50	0.50	1.50	1.50	0.50	3588	313.6
Run16	1.50	0.50	1.50	1.50	0.50	0.50	1.50	0.50	3596	324.8
Run17	0.50	0.50	0.50	0.50	0.50	0.50	0.50	0.50	3086	279.1

Table 4.24 Coefficients for field oil production response surface, case study 3

b0	3661.8
b1	−9.593
b2	169.92
b3	83.30
b4	74.33
b5	33.55
b6	33.22
b7	23.15
b8	99.08

Table 4.25 Coefficients for field gas production response surface, case study 3

b0	319.88
b1	20.62
b2	8.378
b3	7.036
b4	5.783
b5	5.098

4.5 Case study 4

Vast resources of heavy crude oil and the growing world oil demand together with high oil price attract the attention of oil companies to develop new techniques for more oil recovery from these huge resources. Steam-based methods (Cyclic Steam Stimulation, Steam Flooding, Steam-Assisted Gravity Drainage) have been applying in numerous heavy-oil fields of Canada and Venezuela, and are the most advanced of all EOR methods. Among the steam-based methods, steam-assisted gravity drainage (SAGD) is the most frequently used method for heavy oil and bitumen recovery, as the method is claimed to have more recovery factor than other methods; 60% in general, or even as high as 70−80% in some favorable reservoirs [Sheng, 2013]. SAGD is a thermal oil recovery process, invented by Butler in 1982, for recovering viscous heavy oil and bitumen. In SAGD, heavy oil recovery is done by a pair of horizontal wells; steam is continuously injected into the reservoir by an injection well (the upper well) and forms a steam chamber. Steam in the steam chamber releases its latent heat to the reservoir and lowers the viscosity of the oil. Oil is mobilized and drains by gravity toward the production well (the lower well) [Butler, 1982].

The fourth case study illustrates SAGD in a typical Canadian heavy oil.

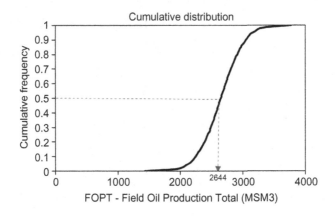

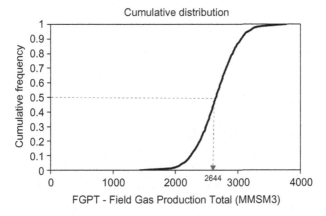

Figure 4.56 Monte Carlo results, case study 3.

4.5.1 Steam assisted gravity drainage in a heavy oil reservoir

In this case study, the effects of different factors on performance of steam assisted gravity drainage (SAGD) into a heavy oil reservoir are examined. The reservoir is at a depth of 500 m with a bulk volume of 50,000 m³ (thickness is estimated 50 m). Cores sampled show that porosity is 0.24 and horizontal and vertical permeability are 4000 md and 2200 md, respectively. Reservoir temperature at 500 m is 10°C and measured oil viscosity at this temperature is 1.6×10^6 cp. Other PVT properties of oil are tabled in Figure 4.57.

Variation of oil viscosity with temperature is correlated as:

$$\mu = 0.000012 \mathrm{Exp}\left(\frac{7275}{T}\right) \tag{4.11}$$

where T is absolute temperature (K).

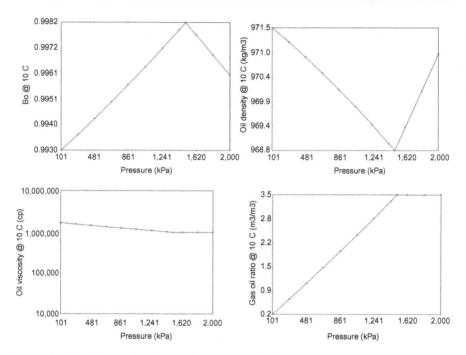

Figure 4.57 PVT properties of oil and gas, case study 4.

Thermal conductivities (in unit J/(m*day*C)) of rock, solid and fluid phases at different temperatures are as follows:

Temp, °C	Rock	Water	Oil	Gas
10	186792.5	26773.58	11207.54	155.66
300	186792.5	26773.58	11207.54	155.66

It is desired to inject steam of 95% quality at 235°C at a rate of 1500 m³/day to enhance oil recovery.

To model the SAGD process, the reservoir is discretized into 5050 (101 × 1 × 50) Cartesian grid cells with uniform porosity and permeability.

Relative permeability curves are generated using Corey correlation as shown in Figure 4.58.

Steam is injected into the reservoir by a well located at a depth of 545 m and oil is produced by a producer well located at 549 m (Figure 4.59).

The SAGD process continues for about 4 years. The study objective is to measure the effects of porosity (POR), vertical permeability (PERMV), horizontal permeability (PERMH) and temperature dependency of relative permeability endpoints (SORW, SORG) on the reservoir performance under the SAGD process.

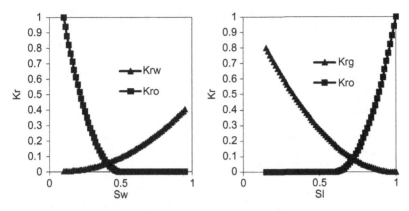

Figure 4.58 Relative permeability curves, case study 4.

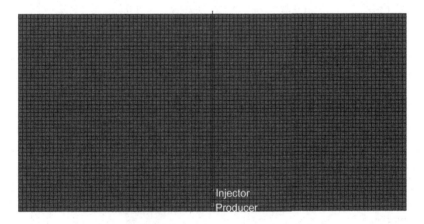

Figure 4.59 Locations of injection and production wells, case study 4.

The reservoir is modeled using Computer Modelling Group, Ltd. STARS thermal reservoir simulation software [Computer Modelling Group, Ltd., 2011].

4.5.2 Experimental design

First, a two-level Plackett-Burman experimental design is applied to figure out the most influential factors (screening). Table 4.26 shows the Plackett-Burman design.

Running the 12 cases using STARS and plotting a Tornado chart indicate that temperature dependency of residual oil saturation in the presence of water and gas (SORW and SORG) have little effect on the reservoir performance (Figures 4.60 to 4.62). SORW is then disregarded in the next design.

In order to investigate more, the next design is conducted using a three-level Box-Behnken design (see Table 4.27). The 25 cases are run by STARS; the result is summarized in Figure 4.63.

Table 4.26 **Plackett-Burman design, case study 4**

Parameter	Low	High
POR	0.22	0.36
PERMH	3000	6000
PERMV	2000	2800
SORW	0.16	0.26
SORG	0.02	0.06

Run number	POR	PERMH	PERMV	SORW	SORG
1	0.22	3000	2000	0.26	0.02
2	0.22	3000	2800	0.16	0.02
3	0.22	3000	2800	0.16	0.06
4	0.22	6000	2000	0.16	0.06
5	0.22	6000	2000	0.26	0.06
6	0.22	6000	2800	0.26	0.02
7	0.36	3000	2000	0.16	0.06
8	0.36	3000	2000	0.26	0.02
9	0.36	3000	2800	0.26	0.06
10	0.36	6000	2000	0.16	0.02
11	0.36	6000	2800	0.16	0.02
12	0.36	6000	2800	0.26	0.06

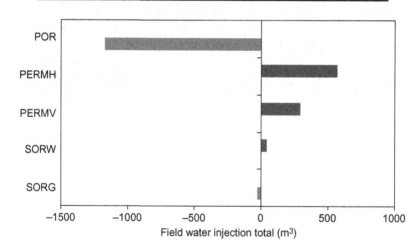

Figure 4.60 Tornado chart for total field water injection, case study 4.

Based on the results of the Box-Behnken design and the linear regression, the relationship between cumulative oil production to the steam-oil ratio (FOPT_SOR_Ratio), and the four important factors are measured as follows:

$$FOPT\_SOR\_Ratio = -228.381 + 354.449 * POR + 0.0274287 * PERMH$$
$$+ 0.117496 * PERMV - 504.453 * SORG$$

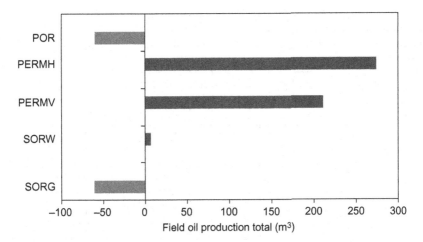

Figure 4.61 Tornado chart for total field oil production, case study 4.

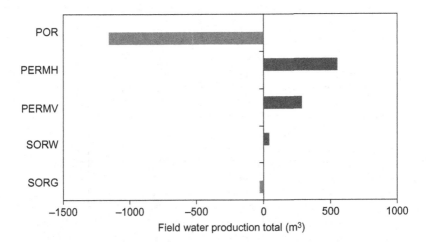

Figure 4.62 Tornado chart for total field water production, case study 4.

The above response surface is then used in a Monte Carlo simulation to generate cumulative distribution and cumulative probability plots of cumulative oil production to steam-oil ratio, shown in Figures 4.64 and 4.65.

4.6   Case study 5

Shale gas reservoirs are unconventional resources and have become important sources of natural gas in the world. It is estimated 32% of the total natural gas resources are in shale formations. Among the top 10 countries with technically

Table 4.27 **Box-Behnken design, case study 4**

Parameter	Lower value	Middle value	Higher value
POR	0.22	0.28	0.34
PERMH (md)	3800	6250	10200
PERMV (md)	1500	2250	3300
SORG	0	0.075	0.15

Run Number	POR	PERMH	PERMV	SORG
1	0.22	3800	2250	0.075
2	0.22	6250	1500	0.075
3	0.22	6250	2250	0
4	0.22	6250	2250	0.15
5	0.22	6250	3300	0.075
6	0.22	10200	2250	0.075
7	0.28	3800	1500	0.075
8	0.28	3800	2250	0
9	0.28	3800	2250	0.15
10	0.28	3800	3300	0.075
11	0.28	6250	1500	0
12	0.28	6250	1500	0.15
13	0.28	6250	2250	0.075
14	0.28	6250	3300	0
15	0.28	6250	3300	0.15
16	0.28	10200	1500	0.075
17	0.28	10200	2250	0
18	0.28	10200	2250	0.15
19	0.28	10200	3300	0.075
20	0.34	3800	2250	0.075
21	0.34	6250	1500	0.075
22	0.34	6250	2250	0
23	0.34	6250	2250	0.15
24	0.34	6250	3300	0.075
25	0.34	10200	2250	0.075

recoverable shale gas resources, China with 1115 tcf is the first, and the United States with 665 tcf is ranked fourth after Algeria [US Energy Information Administration, 2013]. It is estimated that shale gas will supply as much as half the natural gas production in North America by 2020 [Feng, 2013]. These resources cannot be economically developed without enhanced recovery techniques such as multistage hydraulic fracture treatments. Many studies have been conducted to model multistage hydraulic fracture and predict reservoir performance. However, there is still a high level of uncertainty in modeling a shale gas reservoir, including hydraulic fractures.

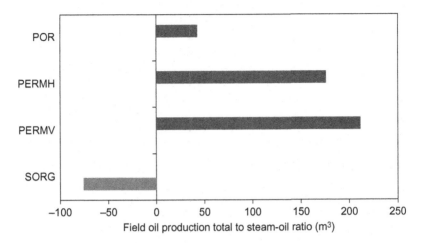

Figure 4.63 Tornado chart for total field oil production to steam-oil ratio, case study 4.

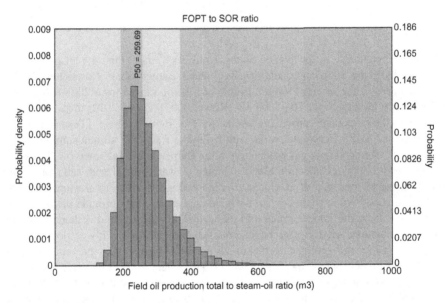

Figure 4.64 Probability density distribution for total field oil production to steam-oil ratio, case study 4.

4.6.1 Barnett shale gas reservoir

The Barnett shale gas reservoir is among the unconventional resources in the United States which have been successfully developed due to hydraulic fracturing treatments. It covers 12950 km^2 and the gas-shale zone is located at a depth of 8000 ft. The shale consists of sedimentary rock made of clay and quartz. Original gas in-place of the Barnett shale gas reservoir is estimated to be 30 tcf [The Perryman Group, 2007]. This gas volume makes Barnett shale gas the largest onshore natural gas field in

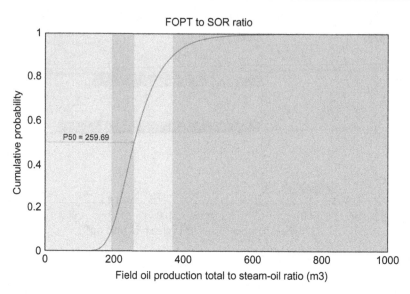

Figure 4.65 Monte Carlo results for total field oil production to steam-oil ratio, case study 4.

Texas and potentially in the United States. Until 2008, more than 4.4 tcf gases have been produced by 10564 producing wells [Powell Barnett Shale Newsletter, 2008]. The thickness of the reservoir within Fort Worth Basin ranges from 200–800 ft with permeability ranging from 70–5000 nD [Reese, 2007]. As of 2012, there were 235 companies that manage producing wells in the Barnett Shale [Texas Railroad Commission, 2012]. Horizontal wells with hydraulic fracturing stimulus are enabling technologies for economic gas production in the Barnett shale gas reservoir. The fracturing process involves injecting a large volume of water mixed with sand and additional chemicals into a well at a pressure high enough to crack or fracture the tight shale formation. When shale cracks, the sand-packed fractures provide pathways for flowing gas from the shale formation to the well. The injected fluid is then recovered from the fractures shortly after being pumped into the well. This is done slowly enough to allow the geologic formation to compress against and prevent the sand from flowing back out of the fractures with the watery fluid. The sand (proppant) remains in the fractures and supports them open. Typically, 500,000 to 1,500,000 gallons of water are mixed with 75,000 to 250,000 pounds of sand (or similar granular substance such as walnut shells or ceramic particles) and a minor amount of additional chemicals (such as hydrochloric acid, sodium chloride, ethylene glycol) and pumped at 45 to 70 bpm (barrels per minute) into the formation [Fisher et al., 2004]. Microseismic technology is then used to monitor the treatments.

4.6.2 Reservoir modeling

Reservoir simulation is an efficient approach to modeling hydraulic fractures and gas flow in a shale gas reservoir. It enables us to evaluate hydraulically fractured horizontal well performance of a shale gas reservoir.

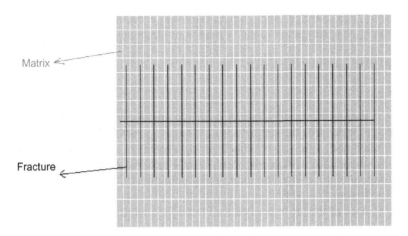

Figure 4.66 Fractures in modeling of Barnett shale gas (X,Y view).

Table 4.28 Barnett shale-gas reservoir information

Parameter	Value
Initial reservoir pressure (psi)	3800
Well bottom hole pressure (psi)	1500
Temperature (°F)	180
Gas viscosity (cP)	0.02
Reservoir depth (ft)	8000
Rock pressure gradient (psi/ft)	0.54
Shale (matrix) permeability (nD)	100
Shale porosity	0.04
Shale compressibility (1/psi)	1E-6
Initial gas saturation	0.70
Fracture height (ft)	300
Fracture spacing (ft)	100

To build the model, a block of Barnett reservoir with one horizontal well is selected. The block length, width and height are 3000 ft, 1500 ft and 300 ft, respectively. The block is discretized into 600 grid cells ($40 \times 15 \times 1$). The horizontal well locates at the center of the block (cell 1,8,1) and covers the whole block length. The well is hydraulically fractured, allowing gas to flow into the well (Figure 4.66). Minimum bottom-hole pressure of the well is set to be 1500 psi. Rock and fluid properties used in the model are given from reference [Yu et al., 2014] and summarized in Table 4.28. In reservoir modeling of Barnett shale gas two regions, one for matrix and another for fracture, are considered and two different rock properties are assigned to the two regions.

As high velocity flow of gas occurs in fractures, non-Darcy fluid flow is considered in the model. The non-Darcy feature we consider here is the Forchheimer correction [Zeng & Grigg, 2005], which takes into account the inertia effects due to high velocity that may occur in high permeability fractures:

$$\frac{dP}{dx} = \left(\frac{\mu}{K.k_r.A}\right)q + \beta.\rho\left(\frac{q}{A}\right)^2 \tag{4.12}$$

where q is the volumetric flow rate, K is the rock permeability, k_r is the relative permeability, A is the area through which flow occurs, μ is the fluid viscosity, ρ is the fluid density and β is the Forchheimer parameter. In reservoir simulators, the Forchheimer parameter β is a user input in the Forchheimer unit: $1F = 1 \text{ atm.s}^2.\text{g}^{-1}$. An alternative unit for β can be obtained by noting that: $1F = 1.01325 \times 10^6 \text{ cm}^{-1}$. Thus, $\beta = 10^7 \text{ cm}^{-1} = 10$ F.

Gas in-place of the block is calculated to be 6158 MMSCF.

4.6.3 Uncertainty parameters

In modeling a shale gas reservoir, properties of created fractures are uncertain; fracture half-length and fracture conductivity (fracture width multiplied by fracture permeability) are two important properties of fractures; see Figure 4.67.

The shale (matrix) permeability and compressibility can play important roles during hydraulic fracturing treatment of a horizontal well.

The Forchheimer parameter β in the non-Darcy flow equation is determined experimentally using a high pressure/high temperature gas flooding device. As there is no experimental study on Barnett shale, we consider the Forchheimer parameter β to be an uncertain parameter. Table 4.29 shows the five uncertain parameters in this case study.

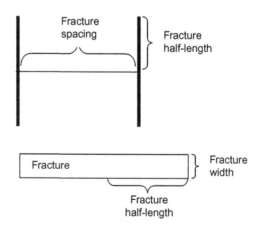

Figure 4.67 Fracture half-length, fracture spacing and fracture width.

Table **4.29** **Uncertain parameters in Barnett shale gas study**

Parameter	Symbol	Low value	Mid value	High value
Fracture conductivity (md.ft)	FCD	0.1	1	10
Fracture half-length (ft)	Xf	300	500	700
Shale permeability (mD)	Km	0.00001	0.0001	0.001
Shale compressibility (1/psi)	CF	1E-6	2E-6	3E-6
Non-Darcy flow coefficient (F)	Beta	20	60	100

Table **4.30** **Plackett-Burman design for Barnet shale gas study**

Run No.	CF (1/psi)	FCD (md.ft)	Km (md)	BETA (F)	Xf (ft)	FGPT (MMSCF)
Run 1	1E-6	0.1	0.001	100	700	1118
Run 2	3E-6	10	0.00001	100	300	187
Run 3	1E-6	10	0.00001	20	300	192
Run 4	3E-6	10	0.00001	100	700	331
Run 5	1E-6	0.1	0.00001	20	300	166
Run 6	2E-6	1	0.0001	60	500	757
Run 7	1E-6	0.1	0.00001	100	700	251
Run 8	3E-6	0.1	0.00001	20	700	274
Run 9	3E-6	0.1	0.001	100	300	959
Run 10	1E-6	10	0.001	20	700	1617
Run 11	3E-6	0.1	0.001	20	300	1104
Run 12	1E-6	10	0.001	100	300	1054
Run 13	3E-6	10	0.001	20	700	1621

4.6.4 Experimental design

Two-level Plackett-Burman design is chosen to discover the most influential parameters on Barnett shale gas performance. Here total gas production after five years (FGPT) is selected as the performance of the reservoir (response function). Considering the above five uncertain parameters and two-level Placket-Burman design, 13 cases should be studied. Table 4.30 shows the 13 cases, their parameters and performances after five years of production. In this design, "Run 6" is a case where the five uncertain factors are at their mid values. The reservoir model of this case is shown in Figure 4.68. Pressure change in the model after five years of gas production is illustrated in Figure 4.69.

Linear regression with interaction between parameters is done to find the relationship between the response function (FGPT) and independent factors.

$$FGPT = b0 + b1 * Km + b2 * MY + b3 * KF + b4 * BETA + b5 * Km * BETA$$

where
$b0 = 740884$, $b1 = 505926$, $b2 = 112773$, $b3 = 77985$, $b4 = -89513$, $b5 = -48620$

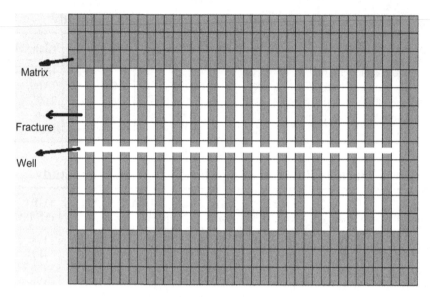

Figure 4.68 Reservoir model of case "Run 6".

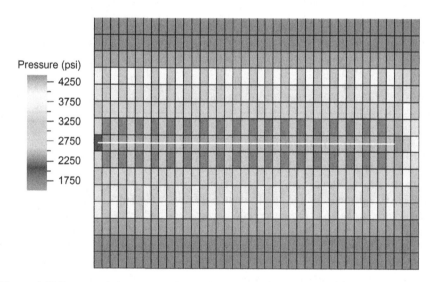

Figure 4.69 Pressure change around the well and fractures after 5 years gas production.

Figure 4.70 shows predicted FGPT versus actual FGPT based on the linear regression. This relationship shows that the most influential factors on FGPT are matrix permeability (Km), fracture half-length (Xf), fracture conductivity (FCD) and non-Darcy flow coefficient (BETA). Figure 4.71 illustrates this dependency as a Pareto chart.

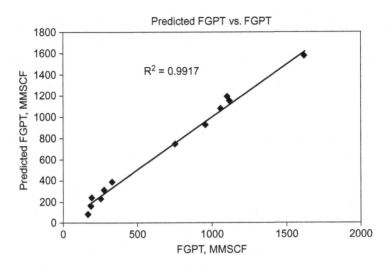

Figure 4.70 Predicted FGPT versus actual FGPT using linear regression.

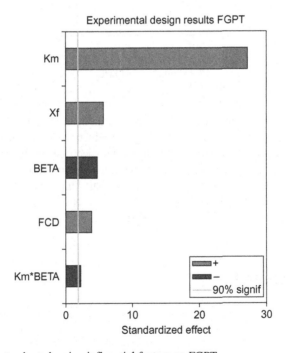

Figure 4.71 Pareto chart showing influential factors on FGPT.

After determination of the most influential factors on field gas production, the experimental design is repeated to predict the field performance after ten years of production. To do that, three-level nonorthogonal hypercube design is selected, with the four uncertain factors of matrix permeability (Km), fracture half-length

Table 4.31 **Three-level nonorthogonal hypercube design for four factors**

Run No.	FCD (md.ft)	Km (md)	Xf (ft)	BETA (F)	FGPT (MMSCF)
Run 1	0.422	0.001000	700	50	2,092
Run 2	0.133	0.000032	700	65	945
Run 3	0.178	0.000075	300	40	1,011
Run 4	0.237	0.000178	500	100	1,247
Run 5	3.162	0.000750	500	30	2,194
Run 6	10.000	0.000042	500	85	1,079
Run 7	1.778	0.000024	700	45	1,166
Run 8	1.334	0.000562	700	95	1,708
Run 9	1.000	0.000100	500	60	1,381
Run 10	2.371	0.000010	300	70	379
Run 11	7.499	0.000316	300	55	1,554
Run 12	5.623	0.000133	700	80	1,445
Run 13	4.217	0.000056	500	20	1,486
Run 14	0.316	0.000013	500	90	586
Run 15	0.100	0.000237	500	35	1,351
Run 16	0.562	0.000422	300	75	1,479
Run 17	0.750	0.000018	500	25	810

(Xf), fracture conductivity (FCD) and non-Darcy flow coefficient (BETA). The design is illustrated in Table 4.31. Total gas production (FGPT) at the end of the tenth year is shown in the last column of the table. The value ranges from 379 MMSCF to 2194 MMSCF. Monte Carlo simulations with 1000 realizations are done to identify the risk in the predicted gas production. To perform Monte Carlo simulations, distribution of each factor should be specified; in this study, triangular distribution is specified for fracture conductivity, matrix permeability and non-Darcy flow coefficient. Distribution of fracture half-length is defined as a normal distribution. The result of 1000 Monte Carlo simulations is depicted in Figure 4.72. The most likely value for FGPT is 1300 MMSCF.

4.7 Case study 6

Today the average worldwide oil recovery factor has increased from 20% in the 1980s to 35% [Eni, 2012]. The increment is a result of reducing the viscosity of the reservoir oil, or by extracting the oil with solvents, or by modifying the interactions between rock and reservoir fluids. These techniques of increasing the amount of recovered crude oil are called *enhanced oil recovery* methods (EOR). EOR methods can obtain the addition of tens of billions of barrels of oil; one percent increase in recovery factor is equivalent to one or two years of world oil production [Eni, 2012]. Therefore, increasing the oil recovery factor is an important factor for conservation of finite energy resources.

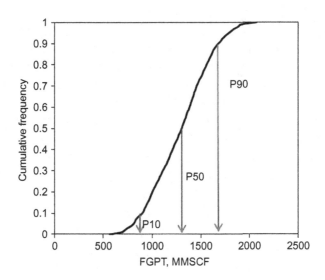

Figure 4.72 Result of Monte Carlo simulations.

In an EOR strategy, improving both microscopic and macroscopic sweep efficiencies should be emphasized by upscaling lab scale results to reservoir scale, and assessing the full field EOR impact by conducting reservoir modeling and simulation. The National IOR Centre of Norway proposed seven tasks for a successful EOR strategy: 1) core scale study (to model the transport mechanisms observed in core scale experiments); 2) nano-scale study (to observe any mineral fluid reactions, such as wettability alteration, at nano-scale); 3) pore scale study (to investigate changes in rock-fluid interactions and wettability at single pores); 4) upscaling (to study the most important parameters from lab scales for describing flow on a reservoir scale); 5) well-to-well tracer injection (to illustrate sweep efficiency at field scale); 6) reservoir simulation; and 7) field-scale evaluation and history matching [The National IOR Centre of Norway, 2014].

Because of the importance of EOR, the sixth case study of this book emphasizes miscible water alternating gas (WAG) injection. In this EOR method, water injection and gas injection are carried out alternately for periods of time to provide better microscopic and sweep efficiencies and reduce gas channeling from injector to producer.

4.7.1 Miscible WAG injection

A $3500 \times 3500 \times 100 \text{ ft}^3$ reservoir at a depth of 8325 ft is considered for the sixth case study [Killough and Kossack, 1987]. The reservoir consists of three geologic layers; thickness, porosity and permeability of each layer are summarized in Table 4.32. Special core analysis (SCAL) is done on core plugs to determine relative permeability data and capillary pressure data. Results are depicted in Figure 4.73. Reservoir fluid is an undersaturated volatile oil. It contains 50% C_1,

Table 4.32 **Reservoir rock data**

Layer	Horizontal permeability (md)	Vertical permeability (md)	Thickness (ft)	Porosity
1	500	50	20	0.3
2	50	50	30	0.3
3	200	25	50	0.3

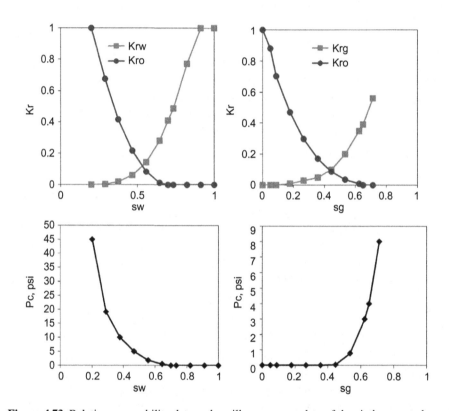

Figure 4.73 Relative permeability data and capillary pressure data of the sixth case study.

3% C_3, 7% C_6, 20% C_{10}, and 15% C_{15}, and 5% C_{20}. Oil, gas and water densities at stock tank conditions are 38.53 lb/ft^3, 0.06864 lb/ft^3 and 62.4 lb/ft^3. Reservoir temperature is 160°F and saturation pressure of reservoir oil at the reservoir temperature is 2302.3 psia. Results of other PVT tests (constant composition expansion test and differential liberation test) are shown in Tables 4.33 and 4.34. An injection well is completed at the top of the reservoir to inject water and gas alternately and oil is produced by a producer completed at the bottom of the reservoir. Injected gas contains 77% C_1, 20% C_3 and 3% C_6. It is desired to inject gas at minimum miscibility pressure to gain maximum microscopic efficiency. Maximum oil rate of the production well is set to be 12000 STB/D.

Table 4.33 Constant Composition Expansion at 160°F

Pressure (psia)	4800	4500	4000	3500	3000	2500	2302.3
Relative oil volume	0.9613	0.9649	0.9715	0.9788	0.9869	0.996	1
Pressure (psia)	2000	1800	1500	1200	1000	500	14.7
Relative oil volume	1.0668	1.1262	1.2508	1.4473	1.6509	2.9317	164.088

Table 4.34 Differential liberation test at 160°F

Pressure (psia)	Oil FVF (rbbl/stb)	Solution gas-oil ratio (scf/stb)	Oil specific gravity	Gas FVF (rcf/scf)	Oil viscosity (cP)
4800	1.2506	572.8	0.5628		0.208
4500	1.2554	572.8	0.5607		0.272
4000	1.2639	572.8	0.5569		0.265
3500	1.2734	572.8	0.5527		0.253
3000	1.2839	572.8	0.5482		0.24
2500	1.2958	572.8	0.5432		0.227
2302.3	1.301	572.8	0.541	0.907	0.214
2000	1.26	479	0.549	0.851	0.208
1800	1.235	421.5	0.5541	0.7352	0.224
1500	1.1997	341.4	0.5617	0.6578	0.234
1200	1.1677	267.7	0.569	0.5418	0.249
1000	1.1478	222.6	0.5738	0.4266	0.264
500	1.1017	117.6	0.5853	0.3508	0.274
14.7	1.0348	0	0.5966	0.1688	0.295

4.7.2 Reservoir modeling

The reservoir described in the previous section is discretized to 147 grid cells ($7 \times 7 \times 3$), Figure 4.74. Reservoir simulation is done using a black-oil simulator. The main part of this reservoir modeling is generating black-oil PVT properties. To accomplish this, PVT tests described in the previous section are modeled using the Peng-Robinson cubic equation of state, Eq. 4.1. A regression technique with the adaptive least-squares algorithm is applied to minimize the difference between experimental data and predicted results of the equation of state, Eq. 4.2. Hydrocarbon-hydrocarbon binary interaction coefficient exponent, critical volume, critical temperature, acentric factor, Ω_A and Ω_B of component C_{20} together with Ω_A and Ω_B of component C_{15} are selected as regression parameters. The results are shown in Figure 4.75. After tuning the equation of state, oil and gas PVT properties for black-oil simulation are generated, Table 4.35.

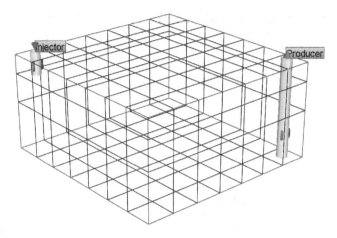

Figure 4.74 Three-dimensional grid cells of the reservoir.

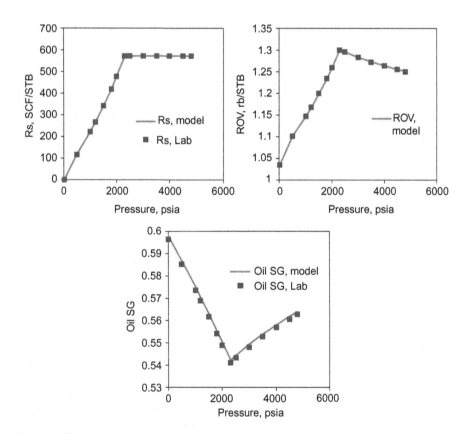

Figure 4.75 Results of PVT modeling.

Table 4.35 **Black-oil PVT data**

Pressure, psia	Gas FVF, rb/MSCF	Gas viscosity, cP
14.7	223.214	0.011
500	5.6022	0.012
1000	2.531	0.013
1200	2.0354	0.014
1500	1.5593	0.016
1800	1.2657	0.018
2000	1.1296	0.019
2302.3	0.9803	0.022
2500	0.9085	0.023
3000	0.7807	0.027
3500	0.6994	0.031
4000	0.643	0.034
4500	0.6017	0.037
4800	0.5817	0.038

Solution Rs, MSCF/STB	Pressure, psia	Oil FVF, rb/STB	Oil viscosity, cP
0	14.7	1.0348	0.31
0.1176	500	1.1017	0.295
0.2226	1000	1.1478	0.274
0.2677	1200	1.1677	0.264
0.3414	1500	1.1997	0.249
0.4215	1800	1.235	0.234
0.479	2000	1.26	0.224
0.5728	2302.3	1.301	0.208
0.5728	3302.3	1.2988	0.235
0.5728	4302.3	1.2966	0.26
0.6341	2500	1.3278	0.2
0.7893	3000	1.3956	0.187
0.9444	3500	1.4634	0.175
1.0995	4000	1.5312	0.167
1.2547	4500	1.5991	0.159
1.3478	4800	1.6398	0.155
1.3478	5500	1.6305	0.168

As the gas is injected at minimum miscibility pressure (MMP), the next stage in the reservoir modeling is evaluating MMP. MMP is the lowest pressure at which miscibility can be achieved at constant temperature and composition. It is a function of the reservoir oil, the composition of the injected gas and the reservoir temperature [Danesh, 1998].

There are three main methods for estimating MMP: laboratory methods (slim tube, rising bubble, contact experiments), phase-behavior calculations based on an EOS, and empirical correlations based on experimental data [Danesh, 1998].

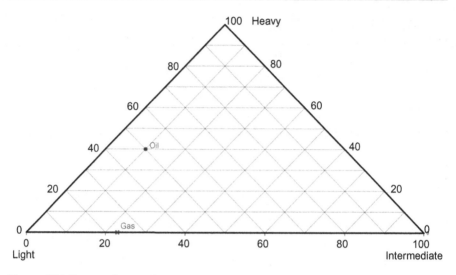

Figure 4.76 Ternary diagram for sixth case study.

In this case study, the MMP is evaluated using two approaches and the results are then compared. The first approach is using a ternary phase diagram of the multicomponent reservoir fluid. The ternary phase diagram is expressed by representing the reservoir fluid with three pseudo-components of light, intermediate and heavy. In this approach the miscibility can be achieved when the compositional path goes through the critical tie line [Danesh, 1998]. C_1 is chosen as light, C_3 and C_6 as intermediate, and C_{10}, C_{15} and C_{20} are selected as heavy (see Figure 4.76).

Then a range of 2300 psia to 3800 psia is chosen for estimating the MMP. At a reservoir temperature of 160°F and each pressure of above range, the tuned Peng-Robinson equation of state is used for drawing a phase envelope on the ternary diagram. At each pressure, a critical tie line is plotted and compared to the criteria of miscibility. The procedure is repeated until miscibility is achieved. Results show that the minimum miscibility pressure is to be 3463 psia (Figure 4.77).

The next approach is analytical calculation of MMP proposed by Wang and Orr [1997]. In this approach, the analytical theory of one-dimensional displacement of oil by multicomponent injection gas is applied to estimate the MMP. The displacement, in the absence of dispersion, is completely controlled by a sequence of key tie lines: those that extend through the original oil composition (initial tie line), injection gas composition (injection tie line) and nc-3 tie lines, known as crossover tie lines. If any one of the key tie lines becomes a critical tie line, multiple contact miscibility develops. Therefore, the MMP is calculated as the lowest pressure at which any one of the key tie lines becomes a critical tie line. This approach estimates MMP to be 2900 psia. Table 4.36 shows oil mole fractions, gas mole fractions and K-values on key tie lines at 2900 psia.

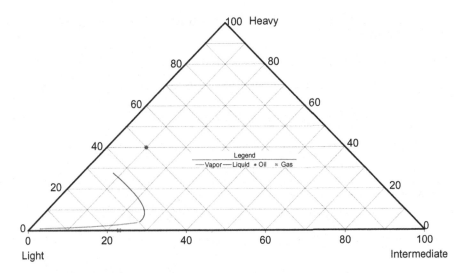

Figure 4.77 Phase diagram at 3463 psia.

The results of a slim tube test show that the minimum miscibility pressure is in the range of 3000 to 3200 psia [Killough and Kossack, 1987].

4.7.3 Uncertain parameters

In this study, six uncertain parameters are selected and we guess they may affect the performance of WAG injection: minimum miscibility pressure (MMP), WAG ratio, gas injection pressure (PG), water injection pressure (PW), WAG cycles (CYCLE), and connate water saturation (SWI). Table 4.37 represents the ranges of these uncertainties.

4.7.4 Experimental design

In order to screen the most influential factors on the performance of WAG injection (oil recovery factor after 10 years injection), two-level Plackett-Burman design is applied (Table 4.38).

Figure 4.78 shows oil recovery factor after 10 years of WAG injection. Minimum and maximum oil recovery factors occur for Run 8 and Run 12, respectively.

Linear regression with interaction between parameters is then used for discovering the most influential parameters. The regression depicts the following relationship between oil recovery factor (FOE) and the most influential parameters:

$$FOE = 0.401 - 0.0513MMP + 0.0473PW + 0.0228SWI$$
$$+ 0.0136MMP.PG + 0.0126PG - 0.067MMP.SWI \tag{4.13}$$

Table 4.36 **Summary of Multiple Contact Miscibility Study at 2900 psia and 160°F**

Oil Mole Fractions on Key Tie Lines

Key tie line	C1	C3	C6	C10	C15	C20
Original oil	0.580117	0.027329	0.059534	0.166886	0.124629	0.041505
Crossover 1	0.574377	0.270487	0.027923	0.067333	0.045566	0.014314
Crossover 2	0.658109	0.241344	0.06236	0.022157	0.01283	0.0032
Crossover 3	0.575341	0.268566	0.081136	0	0.059701	0.015255
Original gas	0.544021	0.276485	0.08748	0	0	0.092013

Gas Mole Fractions on Key Tie Lines

Key tie line	C1	C3	C6	C10	C15	C20
Original oil	0.971233	0.014288	0.008443	0.005229	0.00077	0.000036
Crossover 1	0.777082	0.197218	0.010302	0.011452	0.003514	0.000432
Crossover 2	0.687431	0.230589	0.053993	0.017155	0.00889	0.001943
Crossover 3	0.749428	0.207279	0.035424	0	0.007043	0.000825
Original gas	0.764323	0.201921	0.031444	0	0	0.002311

K-Values on Key Tie Lines

Key tie line	C1	C3	C6	C10	C15	C20
Original oil	1.674202	0.522831	0.141823	0.03133	0.006182	0.000874
Crossover 1	1.352913	0.72912	0.368963	0.170078	0.077122	0.030181
Crossover 2	1.044554	0.955436	0.865828	0.774247	0.692921	0.606978
Crossover 3	1.302581	0.7718	0.436599	1	0.117974	0.054102
Original gas	1.404951	0.730315	0.359441	1	1	0.02512

Table 4.37 **Uncertain parameters and their ranges**

Parameter	Lower value	Middle value	Upper value
MMP (psia)	2900	3200	3500
WAG	2	3.5	5
PG (psia)	4000	4500	4800
PW (psia)	4000	4500	4800
CYCLE (months)	2	3	4
SWI	0.2	0.25	0.3

Figure 4.79 shows predicted FOE versus actual FOE based on Eq. 4.13. A Pareto chart of FOE to discover the most influential parameters is shown in Figure 4.80. The Pareto chart depicts that among the six uncertain parameters, the four parameters of minimum miscibility pressure (MMP), water and gas injection

Table 4.38 **Plackett-Burman design**

Run No.	MMP (psia)	WAG Ratio	PG (psia)	PW (psia)	CYCLE (months)	SWI
Run 1	2900	2	4800	4800	4	0.2
Run 2	3500	5	4000	4800	2	0.2
Run 3	2900	5	4000	4000	2	0.3
Run 4	3500	5	4000	4800	4	0.2
Run 5	2900	2	4000	4000	2	0.2
Run 6	3200	3	4400	4400	3	0.25
Run 7	2900	2	4000	4800	4	0.3
Run 8	3500	2	4000	4000	4	0.3
Run 9	3500	2	4800	4800	2	0.3
Run 10	2900	5	4800	4000	4	0.2
Run 11	3500	2	4800	4000	2	0.2
Run 12	2900	5	4800	4800	2	0.3
Run 13	3500	5	4800	4000	4	0.3

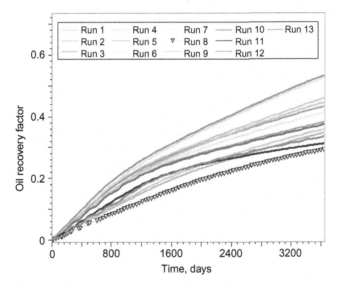

Figure 4.78 Oil recover factor after 10 years WAG injection.

pressures (PW and PG) and connate water saturation (SWI) have the most influence on oil recovery factor in the WAG injection project. Minimum miscibility pressure is the most influential parameter.

For prediction of the performance of WAG injection over the next five years, three-level Box-Behnken design is selected. This design needs to run 27 cases. Table 4.39 shows the design and reservoir performance for each case.

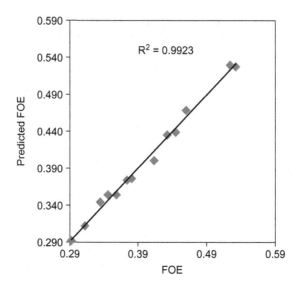

Figure 4.79 Predicted FOE versus actual FOE based on Eq. 4.13.

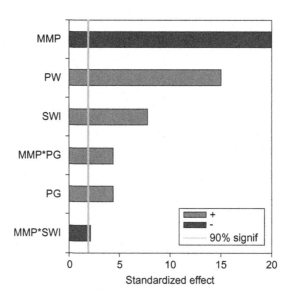

Figure 4.80 Pareto chart for oil recovery factor.

Table **4.39** **Box-Behnken design**

Run No.	MMP psia	PG psia	PW psia	SWI	FOE after 15 years
Case 1	3200	4000	4800	0.25	0.58
Case 2	3500	4000	4400	0.25	0.41
Case 3	3200	4400	4400	0.25	0.52
Case 4	3200	4000	4000	0.25	0.45
Case 5	3200	4000	4400	0.30	0.53
Case 6	2900	4400	4400	0.20	0.54
Case 7	3200	4800	4000	0.25	0.46
Case 8	3200	4400	4000	0.20	0.42
Case 9	2900	4400	4400	0.30	0.61
Case 10	2900	4800	4400	0.25	0.59
Case 11	3200	4800	4400	0.20	0.49
Case 12	2900	4000	4400	0.25	0.58
Case 13	3200	4800	4800	0.25	0.60
Case 14	3500	4400	4400	0.20	0.43
Case 15	3200	4000	4400	0.20	0.48
Case 16	3200	4400	4800	0.30	0.62
Case 17	3500	4400	4400	0.30	0.49
Case 18	3200	4800	4400	0.30	0.56
Case 19	3200	4400	4400	0.25	0.52
Case 20	3500	4400	4000	0.25	0.40
Case 21	3500	4800	4400	0.25	0.47
Case 22	2900	4400	4000	0.25	0.51
Case 23	3200	4400	4000	0.30	0.49
Case 24	3500	4400	4800	0.25	0.53
Case 25	2900	4400	4800	0.25	0.66
Case 26	3200	4400	4800	0.20	0.55
Case 27	3200	4400	4400	0.25	0.52

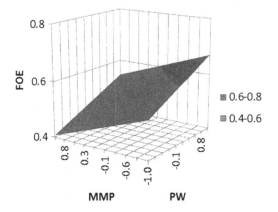

Figure 4.81 Response surface of oil recovery factor.

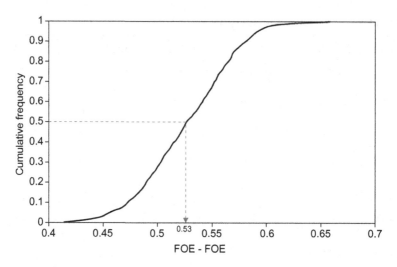

Figure 4.82 Summary of Monte Carlo simulations.

Linear regression with interaction between parameters is selected for response surface modeling. The regression results in the following relationship between reservoir oil recovery factor and the four uncertain parameters:

$$FOE = 0.519 + 0.068PW - 0.064MMP + 0.033SWI + 0.009PG$$
$$+ 0.015MMP.PG \tag{4.14}$$

Response surface in 3-D diagram is shown in Figure 4.81.

Monte Carlo simulations with 1000 realizations are then applied to identify the risk in the predicted oil recovery factor after 15 years of WAG injection. A triangular distribution is selected for MMP, PW, PG, and SWI. Results of the Monte Carlo simulations are shown in Figure 4.82. Based on the Monte Carlo simulation, the most probable value for recovery factor is 0.53.

Appendix
F distribution values

$$*F_{v1,v2} = \frac{\Gamma((v1+v2)/2)(v1/v2)^{v1/2}(x)^{(v1/2)-1}}{\Gamma(v1/2)\Gamma(v2/2)[(v1/v2)x+1]^{(v1+v2)/2}}$$

F Distribution Factor, $F^*_{v1,v1}$

v2 \ v1	1	2	3	4	5	6	7	8	9	10	12	15	20	24	30	40	60	120	∞
1	161.4	199.5	215.7	224.6	230.2	234	236.8	238.9	240.5	241.9	243.9	245.9	248	249.1	250.1	251.1	252.2	253.3	254.3
2	18.51	19	19.16	19.25	19.3	19.33	19.35	19.37	19.38	19.4	19.41	19.43	19.45	19.45	19.46	19.47	19.48	19.49	19.5
3	10.13	9.55	9.28	9.12	9.01	8.94	8.89	8.85	8.81	8.79	8.74	8.7	8.66	8.64	8.62	8.59	8.57	8.55	8.53
4	7.71	6.94	6.59	6.39	6.26	6.16	6.09	6.04	6	5.96	5.91	5.86	5.8	5.77	5.75	5.72	5.69	5.66	5.63
5	6.61	5.79	5.41	5.19	5.05	4.95	4.88	4.82	4.77	4.74	4.68	4.62	4.56	4.53	4.5	4.46	4.43	4.4	4.36
6	5.99	5.14	4.76	4.53	4.39	4.28	4.21	4.15	4.1	4.06	4	3.94	3.87	3.84	3.81	3.77	3.74	3.7	3.67
7	5.59	4.74	4.35	4.12	3.97	3.87	3.79	3.73	3.68	3.64	3.57	3.51	3.44	3.41	3.38	3.34	3.3	3.27	3.23
8	5.32	4.46	4.07	3.84	3.69	3.58	3.5	3.44	3.39	3.35	3.28	3.22	3.15	3.12	3.08	3.04	3.01	2.97	2.93
9	5.12	4.26	3.86	3.63	3.48	3.37	3.29	3.23	3.18	3.14	3.07	3.01	2.94	2.9	2.86	2.83	2.79	2.75	2.71
10	4.96	4.1	3.71	3.48	3.33	3.22	3.14	3.07	3.02	2.98	2.91	2.85	2.77	2.74	2.7	2.66	2.62	2.58	2.54
11	4.84	3.98	3.59	3.36	3.2	3.09	3.01	2.95	2.9	2.85	2.79	2.72	2.65	2.61	2.57	2.53	2.49	2.45	2.4
12	4.75	3.89	3.49	3.26	3.11	3	2.91	2.85	2.8	2.75	2.69	2.62	2.54	2.51	2.47	2.43	2.38	2.34	2.3
13	4.67	3.81	3.41	3.18	3.03	2.92	2.83	2.77	2.71	2.67	2.6	2.53	2.46	2.42	2.38	2.34	2.3	2.25	2.21
14	4.6	3.74	3.34	3.11	2.96	2.85	2.76	2.7	2.65	2.6	2.53	2.46	2.39	2.35	2.31	2.27	2.22	2.18	2.13
15	4.54	3.68	3.29	3.06	2.9	2.79	2.71	2.64	2.59	2.54	2.48	2.4	2.33	2.29	2.25	2.2	2.16	2.11	2.07
16	4.49	3.63	3.24	3.01	2.85	2.74	2.66	2.59	2.54	2.49	2.42	2.35	2.28	2.24	2.19	2.15	2.11	2.06	2.01
17	4.45	3.59	3.2	2.96	2.81	2.7	2.61	2.55	2.49	2.45	2.38	2.31	2.23	2.19	2.15	2.1	2.06	2.01	1.96
18	4.41	3.55	3.16	2.93	2.77	2.66	2.58	2.51	2.46	2.41	2.34	2.27	2.19	2.15	2.11	2.06	2.02	1.97	1.92
19	4.38	3.52	3.13	2.9	2.74	2.63	2.54	2.48	2.42	2.38	2.31	2.23	2.16	2.11	2.07	2.03	1.98	1.93	1.88
20	4.35	3.49	3.1	2.87	2.71	2.6	2.51	2.45	2.39	2.35	2.28	2.2	2.12	2.08	2.04	1.99	1.95	1.9	1.84
21	4.32	3.47	3.07	2.84	2.68	2.57	2.49	2.42	2.37	2.32	2.25	2.18	2.1	2.05	2.01	1.96	1.92	1.87	1.81
22	4.3	3.44	3.05	2.82	2.66	2.55	2.46	2.4	2.34	2.3	2.23	2.15	2.07	2.03	1.98	1.94	1.89	1.84	1.78
23	4.28	3.42	3.03	2.8	2.64	2.53	2.44	2.37	2.32	2.27	2.2	2.13	2.05	2.01	1.96	1.91	1.86	1.81	1.76
24	4.26	3.4	3.01	2.78	2.62	2.51	2.42	2.36	2.3	2.25	2.18	2.11	2.03	1.98	1.94	1.89	1.84	1.79	1.73
25	4.24	3.39	2.99	2.76	2.6	2.49	2.4	2.34	2.28	2.24	2.16	2.09	2.01	1.96	1.92	1.87	1.82	1.77	1.71
26	4.23	3.37	2.98	2.74	2.59	2.47	2.39	2.32	2.27	2.22	2.15	2.07	1.99	1.95	1.9	1.85	1.8	1.75	1.69
27	4.21	3.35	2.96	2.73	2.57	2.46	2.37	2.31	2.25	2.2	2.13	2.06	1.97	1.93	1.88	1.84	1.79	1.73	1.67
28	4.2	3.34	2.95	2.71	2.56	2.45	2.36	2.29	2.24	2.19	2.12	2.04	1.96	1.91	1.87	1.82	1.77	1.71	1.65
29	4.18	3.33	2.93	2.7	2.55	2.43	2.35	2.28	2.22	2.18	2.1	2.03	1.94	1.9	1.85	1.81	1.75	1.7	1.64
30	4.17	3.32	2.92	2.69	2.53	2.42	2.33	2.27	2.21	2.16	2.09	2.01	1.93	1.89	1.84	1.79	1.74	1.68	1.62
40	4.08	3.23	2.84	2.61	2.45	2.34	2.25	2.18	2.12	2.08	2	1.92	1.84	1.79	1.74	1.69	1.64	1.58	1.51
60	4	3.15	2.76	2.53	2.37	2.25	2.17	2.1	2.04	1.99	1.92	1.84	1.75	1.7	1.65	1.59	1.53	1.47	1.39
120	3.92	3.07	2.68	2.45	2.29	2.17	2.09	2.02	1.96	1.91	1.83	1.75	1.66	1.61	1.55	1.5	1.43	1.35	1.25
∞	3.84	3	2.6	2.37	2.21	2.1	2.01	1.94	1.88	1.83	1.75	1.67	1.57	1.52	1.46	1.39	1.32	1.22	1

References

Ahmed, T., Meehan, T., 2011. Advanced Reservoir Management and Engineering. Gulf Professional Publishing Company, Houston, Texas.

Amaefule, J.O., Altunbay, M., Tiab, J., Kersey, D.G., Keelan, D.K. Enhanced Reservoir Description: Using Core and Log Data to Identify Hydraulic (Flow) Units and Predict Permeability in Uncored Intervals/Wells. Paper SPE 26436 presented at the 68th Annual Technical Conference and Exhibition of the Society of Petroleum Engineers held in Houston. Texas, 3−6 October 1993.

Amyx, J.W., Bass, J.D., Whiting, R., 1960. Petroleum Reservoir Engineering: Physical Properties. McGraw-Hill, New York, NY.

Antony, J., 2003. Design of Experiments for Engineers and Scientists. Butterworth-Heinemann, Oxford.

Aziz Khalid, D.L., Tchelepi, H.A., 2005. Notes on Reservoir Simulation. Stanford University, Stanford, CA.

Baldwin, D.E., 1969. A Monte Carlo Model for Pressure Transient Analysis. Paper SPE 2568 presented at 1969 SPE Annual Meeting, Denver, 28 September−1 October.

Bear, J., 1972. Dynamics of Fluids in Porous Media. Dover Publications.

Butler, R.M., 1982. Method for Continuously Producing Viscous Hydrocarbons by Gravity Drainage While Injecting Heated Fluids, US Patent No. 4,344,485.

Caers, J., 2005. Petroleum Geostatistics. Society of Petroleum Engineers, Richardson, Texas.

Chen, Z., 2007. Reservoir Simulation Mathematical Techniques in Oil Recovery. Society for Industrial and Applied Mathematics, SIAM.

Chen, H.-S., Stadtherr, M.A., 1981. A modification of powell's dogleg method for solving systems of nonlinear equations. Comp. Chem. Eng. 5 (3), 143−150.

Cheong, Y.P., Gupta, R., 2005. Experimental design and analysis methods for assessing volumetric uncertainties. SPE J. 324−335.

Christiansen, R.L., 2001. Two-Phase Flow Through Porous Media: Theory, Art and Reality of Relative Permeability and Capillary Pressure. Colorado KNQ Engineering.

Chueh, P.L., Prausnitz, J.M., 1967. Vapor-Liquid Equilibria at high pressures: calculation of partial molar volumes in nonpolar liquid mixtures. AIChE J. 13 (6), 1099−1107.

Coats, K.H., Smart, G.T., 1986. Application of a regression-based EOS PVT program to laboratory data. SPE Reservoir Eng. 1 (3), 277−299.

Computer Modelling Group, Ltd. (CMG), 2011. STARS Advanced Processes and Thermal Reservoir Simulator. Available at: <www.cmgl.ca>.

Corre, B., Thore, P., deFeraudy, V., Vincent, G. Integrated Uncertainty Assessment for Project Evaluation and Risk Analysis, paper SPE 65205, presented at SPE European Petroleum Conference, Paris, France, 24−25 October 2000.

Dake, L.P., 1978. Fundamentals of Reservoir Engineering. Elsevier, Amsterdam.

Danesh, A., 1998. PVT and Phase Behavior of Petroleum Reservoir Fluids. Elsevier, Oxford.

Davis, J.C., 2002. Statistics and Data Analysis in Geology. John Wiley and Sons, New York.

Dennis Jr., J.E., Gay, D.M., Welsch, R.E., 1981. An adaptive nonlinear least-squares algorithm. ACM Trans. Math. Software. 7 (3), 348−368.

Deutsch, C.V., Journel, A.G., 1992. GSLIB: Geostatistical Software Library and User's Guide. Oxford University Press, Oxford.

Dietrich, P., Helmig, R., Sauter, M., Hotzl, H., Kongeter, J., Teutsch, G., 2005. Flow and Transport in Fractured Porous Media. Springer.

Dujardin, B.O., Matringe S.F. Practical Assisted History Matching and Probabilistic Forecasting Procedure: A West Africa Case Study, SPE 146292, SPE Annual Technical Conference and Exhibition, Denver, Colorado, USA, 30 October−2 November 2011.

ECLIPSE, a Black-Oil Reservoir Simulator, Schlumberger Company, 2011. Available from: <www.slb.com>.

Eni Innovation & Technology, 2012. Available from: <www.eni.com>.

Ertekin, T., Abou-Kassem, J., King, G.R., 2001. Basic Applied Reservoir Simulation. Turgay; SPE Testbook Series.

Fanchi, J.R., 2010. Integrated Reservoir Asset Management. Gulf Professional Publishing Company, Houston, Texas.

Feng, Z., 2013. The Impact of the Changing Global Energy Map on Geopolitics of the World. China-United States Exchange Foundation. Retrieved 15.05.2013.

Fisher, M.K., Heinze, J.R., Harris, C.D., Davidson, B.M., Wright, C.A., Dunn, K.P., 2004. Optimizing horizontal completion techniques in the Barnett Shale using microseismic fracture mapping: Proceedings of the Society of Petroleum Engineers Annual Technical Conference, Houston, Texas, SPE Paper 90051.

Friedmann, F., Chawathe, A., Larue, D.K. Assessing Uncertainty in Channelized Reservoirs Using Experimental Designs, paper SPE 71622 presented at the 2001 Annual Technical Conference and Exhibition, New Orleans, 30 September−3 October.

Gad, S.C., 2006. Statistics and Experimental Design for Toxicologists and Pharmacologists. Taylor & Francis.

Gilman, J.R., Brickey, R.T., Red, M.M., Monte Carlo Techniques for Evaluating Producing Properties, paper SPE 39925 presented at the 1998 SPE Rocky Mountain Regional Low Permeability Reservoirs Symposium, Denver, 5−8 April.

Harris, D.G., 1975. The role of geology in reservoir simulation studies. J. Petroleum Technol. 625−632.

IHS Energy, IHS WellTest software, 2014. Available from: <https://www.ihs.com/products/welltest-oil-reserve-pta-software.html>.

Islam, M.R., Mousavizadegan, S.H., Shabbir, M., Abou-Kassem, J., 2007. Advanced Petroleum Reservoir Simulation. McGraw Hill Publishing, New York.

Kazemi, H., 1976. Numerical simulation of water-oil flow in naturally fractured reservoirs. SPE 5719, SPEJ. 317−326.

Kelkar, M., Perez, G., 2002. Applied Geostatistics for Reservoir Characterization. Society of Petroleum Engineers, Richardson, Texas.

Khosravi, M., Rostami, B., Fatemi, 2012. Uncertainty analysis of a fractured reservoir's performance: a case study. Oil Gas Sci. Technol. 67 (3), 423−433.

Killough J.E. 9th SPE Comparative Solution Project: A Reexamination of Black-Oil Simulation, Paper SPE 29110 presented at the 13th SPE Symposium on Reservoir Simulation held in San Antonio, TX, 12−15 February 1995.

Killough, J.E., Kossack, C.A. Fifth SPE Comparative Solution Project: Evaluation of Miscible Flood Simulators, Society of Petroleum Engineers 16000 paper presented at the Ninth SPE Symposium on Reservoir Simulation held in San Antonio, TX, 1−4 February 1987.

King, P.R., 1996. Upscaling permeability: error analysis for renormalization. Trans. Porous Media. 23, 337−354.

Kleppe, J., 2004. Lecture Notes on Reservoir Simulation. Course presented at Department of Petroleum Engineering and Applied Geophysics, Norwegian University of Science and Technology.

Lake, L.W., Carroll Jr., H.B., 1986. Reservoir Characterization. Academic Press, Orlando.

Lazić, Z.R., 2004. Design of Experiments in Chemical Engineering. Wiley-VCH, New York.

Mattax, C.C., Dalton, R.L., 1989. Reservoir Simulation, SPE Monograph Series.

Middle East Well Evaluation Review, 1997. Carbonates: The Inside Story, Middle East & Asia Reservoir Review, 18.

Montgomery, D.C., 2001. Design and Analysis of Experiments. John Wiley and Sons Inc., New York.

Murtha, J.A. Infill Drilling in the Clinton: Monte Carlo Techniques Applied to the Material Balance Equation, paper SPE 17068 presented at the 1987 SPE Eastern Regional Meeting, Pittsburgh, 21−23 October.

Murtha, J.A. Incorporating Historical Data in Monte Carlo Simulation, paper SPE 26245 presented at the 1993 SPE Petroleum Computer Conference, New Orleans, 11−14 July.

The National IOR Centre of Norway, 2014. Available from: <www.uis.no/getfile.php/ Admin-Kategorivisningsmaler/NationalIORCentreFinalNRF.pdf>.

NIST Information Technology Laboratory, 2012. NIST/SEMATECH e-Handbook of Statistical Methods. Available from: <www.itl.nist.gov/div898/handbook/pri/section3/ pri335.htm>.

Odeh, A., 1982. An overview of mathematical modeling of the behavior of hydrocarbon reservoirs. SIAM Rev. 24 (3), 263.

OPEC Share of World Oil Reserves 2013. OPEC, 2014. Available at: <http://www.opec.org/ opec_web/en/data_graphs/330.htm>.

Pawar, R.J., Tartakovsky, D.M. Propagation of measurement errors in reservoir modeling, Proceeding of the XIII International Conference on Computational Methods in Water Resources, Calgary, Canada, 25−29 June 2000.

Peake, W.T., Abadah, M., Skander, L., Uncertainty Assessment Using Experimental Design: Minagish Oolite Reservoir, SPE 91820, Reservoir Simulation Symposium, Houston TX, 31 January−2 February 2005.

Peng, D.-Y., Robinson, D.B., 1976. A new two-constant equation of state. Ind. Eng. Chem. Fundam. 15, 59−64.

Peng, D.-Y., Robinson, D.B., 1977. A rigorous method for predicting the critical properties of multicomponent systems from an equation of state. AIChE J. 23, 137−144.

Portella, R.C.M., Salomao, M.C., Blauth, M., Duarte, R.L.B., Uncertainty Quantification to Evaluate the Value of Information in a Deepwater Reservoir, paper SPE 79707 presented at the 2003 SPE Reservoir Simulation Symposium, Houston, 3−5 February.

Powell Barnett Shale Newsletter, 2008. A History and Overview of the Barnett Shale. Available at: <http://www.barnettshalenews.com>.

Reese, J.L., 2007. Simulating Gas Production from Hydraulic Fracture Networks: A Case Study of the Barnett Shale, M.S.E., The University of Texas at Austin.

Rietz, D., Palke, M., 2001. History matching helps validate reservoir simulation models. Oil Gas J. December 24. Available at: <http://www.ogj.com/articles/print/volume-99/issue-52/drilling-production/history-matching-helps-validate-reservoir-simulation-models. html>.

Santner, T.J., Williams, B.J., Notz, W.I., 2003. The Design and Analysis of Computer Experiments. Springer-Verlag, New York.

Sarma, D.D., 2009. Geostatistics with Application in Earth Sciences. Springer-Verlag, New York.

Satter, A., Thakur, G.C., 1994. Integrated Petroleum Reservoir Management: A Team Approach. PennWell Publishing Company, Tulsa, OK.

Schulze-Riegert, R., Ghedan, S. Modern Techniques for History Matching, 9th International Forum on Reservoir Simulation, Abu Dhabi, 2007.

Selley, R.C., 1998. Elements of Petroleum Geology. Academic Press, San Diego, CA.

Sheng, J.J., 2013. Enhanced Oil Recover Field Case Studies. Elsevier.

Shepard, D. A two-dimensional interpolation function for irregularly-spaced data, Proceedings of the ACM National Conference, 1968.

Sobol, I.M., 1974. The Monte Carlo Method. The University of Chicago Press.

Steppan, D.D., Werner, J., Yeater, R.P., 1998. Essential Regression and Experimental Design for Chemists and Engineers. Available from: <http://www.jowerner.homepage.t-online.de>.

Texas Railroad Commission, 2012. Available from: <http://www.rrc.state.tx.us>.

The Perryman Group (TPG), 2007. Barnett Shale Economic Impact Study. Available from: <www.perrymangroup.com>.

Tiab, D., Donaldson, E.C., 2004. Petrophysics: Theory and Practice of Measuring Reservoir Rock and Fluid Transport Properties. Gulf Professional Company, Houston, TX.

Yu, W., Gao, B., Sepehrnoori, K., 2014. Numerical Study of the Impact of Complex Fracture Patterns on Well Performance in Shale Gas Reservoirs. J. Pet Sci. Res. (JPSR). 3 (2).

Wang, Y., Orr Jr., F.M., 1997. Analytical calculation of minimum miscibility pressure. Fluid Phase Equilib. J. 139, 101.

Warren, J.E., Root, P.J., 1963. The behavior of naturally fractured reservoirs. SPE J. 3 (3), 245–255, SPE-426-PA.

White, C.D., Royer, S.A. Experimental Design as a Framework for Reservoir Studies, paper SPE 79676 presented at the 2003 SPE Reservoir Simulation Symposium, Houston, 3–5 February.

White, C.D., Willis, B.J., Narayanan, K., Dutton, S.P., 2001. Identifying and estimating significant geologic parameters with experimental design. SPEJ. 311.

Wiggins, M.L., Zhang, X. Using PC's and Monte Carlo Simulation to Assess Risks in Workover Evaluations, paper SPE 26243 presented at the 1993 SPE Petroleum Computer Conference, New Orleans, 11–14 July.

WinProp, Phase-Behavior and Fluid Property Program, <www.cmgl.ca>.

Zee Ma, Y., La Pointe, P., 2011. Uncertainty Analysis and Reservoir Modeling: Developing and Managing Assets in an Uncertain World, AAPG: Tulsa, OK.

Zhang, G., 2003. Estimating Uncertainties in Integrated Reservoir Studies. PhD Thesis, Texas A&M University.

Zhangxin, C., 2007. Reservoir Simulation Mathematical Techniques in Oil Recovery, Society for Industrial and Applied Mathematics, SIAM.

Index

A

Adjusted coefficient of determination, 91
Analogy, 9
Analytical models, 45
Arithmetic mean, 25

B

Black-oil, 40, 49, 64
Blocking, 77, 78
Box-Behnken, 79, 87, 88, 105, 106, 145,
    146, 148
Buckley-Leverett problem, 47
Buildup test, 3

C

Cap rock, 1
Capillary pressure, 10, 24, 37, 39, 40, 42,
    62, 111, 112, 118
Carbonate, 1, 4, 109
Coefficient of determination, 91
Conductivity, 153, 156
Confidence interval, 91
Constant composition expansion, 35
Constant volume depletion, 35
Continuity equation, 47, 49, 58, 59
Core data, 10, 11
Corey correlation, 102, 144

D

Darcy's law, 3, 10, 57, 58
Data gathering, 9, 62
Decline curve analysis, 44
Differential liberation, 35
Diffusion coefficient, 60
Digenesis processes, 1, 12
Discretization, 58
Drawdown test, 3
Dual porosity model, 116

E

ECLIPSE, 120, 134, 138, 172
Enhanced oil recovery, ix, 51, 62

Equations of state, 36, 37
Errors, 6, 17, 22, 57, 58
Experimental design, ix, 66, 75, 78, 83, 84,
    103, 105, 145

F

F ratio, 91
Facies, 10
Flow Zone Indicator, 40
Formation volume factor, 34
Fractional factorial, 79, 85, 87, 138, 141
Fractures, 12, 109, 118
Full factorial, 79

G

Gas injection, 9, 51
Gas-oil ratio, 34
Gaussian, 14, 17
Geological, ix, 1, 4, 10, 11, 12, 24, 25, 61,
    62, 134
Geological model, ix, 1
Geometric mean, 25
Geophysical, ix, 1, 4, 10, 11, 61, 62
Geostatistics, 12, 99

H

Harmonic mean, 25
Heterogeneity, 4, 5, 109, 134
History-matching, ix, 9, 64, 66, 67

I

Initialization, 40
Interfacial tension, 63

K

Kriging, 17

L

Least-squares technique, 90
Leverett J-function, 40, 111
Lithology, 12, 24, 25, 109
Log-normal distribution, 71

M

Material balance, 57, 58, 72, 119, 120
Mathematical methods, 9
Mean square error, 91
Method of characteristics, 47
Monte Carlo, 66, 72, 73, 76, 105, 127, 131, 132, 133, 140, 143, 147, 150, 171, 172, 173, 174
Multicolinearity, 91

N

Natural depletion, 4, 49
Normal distribution, 70, 105
Nugget, 13, 21
Numerical dispersion, 60, 61

O

One-parameter-at-a-time, 76
Original oil in place, 3

P

Parachor, 36
Pareto, 79, 103, 120, 122, 123, 124, 125, 138, 140
Peng-Robinson, 36
Permeability, 2, 3, 9, 10, 12, 24, 26, 28, 37, 39, 46, 58, 62, 64, 65, 66, 72, 77, 84, 91, 102, 103, 108, 109, 111, 112, 116, 118, 119, 121, 134, 136, 137, 143, 144, 145
Petroleum traps, 1
Pinchout, 11
Plackett—Burman, 79
Porosity, 2, 64, 99, 134
Power averaging, 25
Pressure equation solution, 25
Pressure transient data, 26
Primary recovery, 4
PUNQ, 134
PVT, 10, 28, 30, 35, 36, 37, 52, 55, 58, 62, 63, 101, 114, 115, 134, 135, 143, 144, 171

R

Randomization, 77, 78
Regression, 37, 90, 105, 115, 116, 122, 138, 146

Regression mean square error, 91
Regression Sum of Squares, 90
Renormalization, 25
Replication, 77
Reservoir characterization, 6, 10, 11, 12, 61, 63
Reservoir model, 6, 9, 11, 30, 63, 64, 102, 116, 120, 134
Response surface, 87, 90, 105, 108, 142, 147
Routine core tests, 24

S

SAGD, 143, 144
Sandstone, 1, 4
Screening, 86, 99
Secondary recovery, 4
Sedimentary rock, 1
Sedimentation process, 12
Sensitivity analysis, 96, 97, 174
Separator test, 35
Shale, 1, 9, 24, 99, 109, 149, 152
Shape factor, 77, 116, 118
Simulator, ix, 1, 120, 134, 138
Source rocks, 1
Special core analysis, 10
Sum of Squared Errors, 90

T

Taylor series, 59
Tornado chart, 79
Total sum of squares, 90
Truncation error, 60

U

Uncertain parameters, 78, 79, 103, 120, 128, 137
Uncertainties, ix, 6, 57, 61, 62, 63, 75, 134
Upscaling, 24, 25

V

Variance inflation factors, 91
Variogram, 12, 18, 21

W

Well logging data, 25

Printed in the United States
By Bookmasters